Qualitative Organic Analysis

Revised Edition

Qualitative Organic Analysis

Revised Edition

William Kemp

Heriot-Watt University, Edinburgh

McGRAW-HILL Book Company (UK) Limited

London · New York · St Louis · San Francisco · Auckland
Bogotá · Guatemala · Hamburg · Johannesburg · Lisbon
Madrid · Mexico · Montreal · New Delhi · Panama · Paris
San Juan · São Paulo · Singapore · Sydney · Tokyo · Toronto

Published by

McGRAW-HILL Book Company (UK) Limited
MAIDENHEAD · BERKSHIRE · ENGLAND

Kemp, William
    Qualitative organic analysis. — Revised ed.
    1. Chemistry, Analytic — Qualitative — Laboratory manuals
    2. Chemistry, Organic — Laboratory manuals
    I. Title
    547′.34′0028   QD271.4   78–41217

    ISBN 0–07–084092–X

567890 W.C. & S. 83210

PRINTED AND BOUND IN GREAT BRITAIN

To Ian, Gillian, and Derek

Contents

Preface

The advent of spectroscopic techniques in organic chemistry has changed drastically the methods used to identify organic compounds. The practising organic chemist no longer relies solely on chemical reactivity in detecting functional groups within a molecule, but makes free use of infrared, ultraviolet, nuclear magnetic resonance, and mass spectrometric data to qualify and supplement the deductions he has made by examining the compound in the test-tube.

These two approaches to the identification of a compound, the chemical and the spectroscopic methods, should not be considered in isolation and are not mutually exclusive; each must be given due weight. But the balance between them is one of the most intriguing aspects of organic chemistry: it is the aim of this book to help the student to achieve the proper balance.

Chemical and spectroscopic information about an unknown compound is sought in an integrated way, and each technique continuously qualifies and supplements the other; the strength of one over-rides the weaknesses and ambiguities of the other. From a pedagogic point of view, the resulting identification scheme has some disadvantages. In particular the identification of very simple compounds (lower aliphatic ketones, monohydric phenols, etc.) appears inordinately elaborate.

The major advantage of this scheme, however, is that it can be applied to the identification of the simplest compounds, and then applied without change to difunctional or trifunctional compounds. Most important of all, the same basic approach is used by the practising organic chemist to identify the functional groups within the most complex of organic molecules. In this respect, the scheme constitutes a sound grounding in qualitative organic analysis, none of which has to be 'unlearnt' in later practice.

The principal criticism of most other texts on qualitative organic analysis is that, on the whole, they tend toward the reverse: that is, they make it easy to identify *simple* organic compounds (often by means of 'specific' colour reactions, etc.) but these techniques fail completely when applied to more complex molecules. Another criticism is that spectroscopic information is not *integrated* into the identification process, but is kept separate from the chemical examination, as an addendum to it.

This book is designed to be used on the laboratory bench by the student. It is assumed that he has access to the infrared spectrum of the unknown organic compound (either recorded by himself or by a technician) and that he has had introductory lectures on the theory and practice of infrared spectroscopy.

Chapter 1 describes a number of general tests which are designed to elicit a fairly comprehensive picture of the chemical behaviour of the unknown compound. These should be carried out fully on all compounds.

In chapter 2 the compound is classified initially on the basis of the elements present, and then further sub-classified into functional classes by means of its chemical and spectroscopic (infrared) properties. Chapter 2 is therefore a systematic hunt for functional groups, but, in multifunctional compounds, not all the functional groups may be initially identified. Rather will the *dominant* functional group(s) be found. For example, hydroxy ketones may be identified simply as ketones; nitro carboxylic acids identified as acids; keto–enol mixtures will in most instances be identified either as the keto *or* the enol form, whichever predominates.

Chapter 3 goes further by examining the infrared spectrum systematically, looking for (a) functional groups whose properties were submerged under some other functional group present in the molecule (e.g., the OH group of hydroxy ketones) and (b) confirmation and cross-checks on those functional groups identified in chapter 2.

Chapter 4 describes the preparation of standard derivatives for a large range of compounds. The list is not exhaustive, but covers all but the most exotic derivatives. Tables of melting points, etc., of derivatives are included at the end of this chapter.

Final identification of the unknown rests on preparing one (or occasionally more than one) derivative, whose physical properties match those recorded in the literature. The student may be permitted in some cases to identify his unknown finally by comparing its infrared spectrum with that of an authentic sample; this is an extremely reliable method, but since authentic samples may not always be available, identification by derivatives as a general method should be duly emphasized.

Not all students will have access to ultraviolet, n.m.r., or mass spectrometers, and these techniques are therefore not incorporated as essentials into the identification process. Their importance, however, demands their inclusion in any text on qualitative analysis, and the discussions given assume that the student has had a course of introduction to each subject.

Where ultraviolet spectroscopy can be of invaluable assistance (as in α,β-unsaturated ketones, or polynuclear aromatic hydrocarbons, etc.) the student is referred to the relevant sections, particularly in chapter 6. It is likely that within a few years n.m.r. spectroscopy will become the commonplace that infrared now is, and chapter 5 summarizes the features which are most important in the identification of an unknown compound. Where a spectrometer is not available, n.m.r. data may be supplied to the student from published information. The same is true of mass spectrometry, and if the student is given mass spectral data for his unknown compound, he can refer to chapter 7 for an outline of the particular points in the mass spectrum which are of greatest assistance in elementary identifications.

Chapter 8 discusses more fully certain of the chemical tests involved, and it is to be hoped that the student will refer to this chapter frequently; the basis of many of the tests is described, together with any interesting mechanistic features. There are good arguments for dispersing these cogent discussions throughout the book, but this has the great disadvantage of overburdening the text with side-comments and diversions. As this would detract from the concise presentation of the analysis scheme, it has been decided to present most of this material as a separate chapter, although a good number of mechanistic equations are nevertheless given along with the descriptions of experimental details throughout the book.

At the back of the book is a page of printed name tabs for the main sections of the book; these can be cut out and glued on to the appropriate pages of the book for easy access.

No scheme of analysis can cover all the variations in functional group behaviour brought about when two or more functional groups are present in a molecule, and the relative positions in the molecule of any two functions is an additional variable which will often profoundly affect the way in which they behave.

This scheme of identification is an attempt to bring closer together the teaching of undergraduates and the workings of the research chemist. To do this, it has been necessary to present as 'generalities' many observations for which exceptions to the rule are known. Finding, recording, and studying these exceptions should be an essential and fascinating part of qualitative organic analysis.

In compiling this book, the author has been pleased to lean heavily on the experience and work of innumerable other chemists, but any deficiencies in this book the author claims unto himself. He would be grateful to have them brought to his attention.

WILLIAM KEMP

Preface to revised edition

The main ethos of this book, when first published, was that spectroscopic methods of identifying organic compounds had to be complemented by 'chemical' evidence, and in this revised edition that concept has been retained.

A case might be made for omitting certain of the test-tube evaluations given in chapter 1, and without doubt the more experienced student will perform fewer of these than the beginning student. For that student coming to qualitative organic analysis for the first time however, there is a strong argument for his carrying out all of the chemical tests systematically—otherwise how will he become experienced?

The mental processes involved in working out the structure of an organic compound are iterative—indeed this is part of the enjoyable challenge—and the mind jumps from i.r. to n.m.r. to chemical evidence and back to i.r. again; but a book must be organized linearly. The Analysis Summary (page xv) shows only *one* possible sequence of operations, and students should eventually aim to organize themselves to use a sequence that suits them best.

Many more melting points have been added to the tables of physical constants, and δ units have replaced τ in n.m.r. spectroscopy. More IUPAC nomenclature has been used for naming the compounds listed, but common names still have a correct place in organic nomenclature and many are retained. Abbreviations such as Me, Et, Ph are used together with the corresponding CH_3, C_2H_5, C_6H_5, etc. so that students can become familiar with both systems; IUPAC radicofunctional nomenclature has been retained for many ketones.

The aim of this book never has been merely to identify organic compounds, but rather to show qualitative organic analysis is a vehicle for *learning about* organic compounds—their chemistry and their spectroscopy. It should be far better to travel hopefully than to arrive.

WILLIAM KEMP

Analysis Summary

(You may equally well choose to carry out (b) before (a))

(a) Perform all of the general chemical tests (chapter 1). Use the results of the sodium fusion test to assign the compound to one of the classes listed on page 9 (chapter 2).

(b) Record the infrared spectrum (chapter 3).

(c) From a *simultaneous* interpretation of the results of (a) and (b) deduce the probable nature of the functional group(s) in the unknown compound: confirm systematically from the given chemical and spectroscopic behaviour of each functional class (chapter 2, pages 11–38).

(d) Examine the infrared spectrum carefully for the presence of functions which have few distinguishing chemical reactions (chapter 3).

(e) Knowing the functional group(s) present, consult the tables of physical constants (chapter 4, pages 86–124) and tentatively identify the compound.

(f) Use all other spectroscopic information available to you (n.m.r., chapter 5; ultraviolet, chapter 6; mass, chapter 7) to modify or confirm your conclusions.

(g) Prepare suitable derivatives (chapter 4) and compare their physical properties with those listed in the tables.

(h) Make a definite identification of the compound.

(i) Present a complete report, and submit labelled derivatives and annotated spectra for inspection.

1

Chemical Examination of an Unknown Compound

Examine the substance according to *all* the screening tests 1–13 below: perform each test as described and record your observation *immediately* in your notebook. Special points to look for in each test are given, and you should pay careful attention to these.

For solid unknowns, the sample should be recrystallized to constant m.p. to ensure purity, unless you are instructed to the contrary; by means of small-scale trials in the test-tube, choose a solvent which gives good crystalline form and high recovery.

All liquids might profitably be examined by gas chromatography (if facilities are available) to assess their purity. When the sample is contaminated with substantial amounts of impurity (say 5–10 per cent), purification should be attempted.

Simple distillation will only remove major impurities which have boiling points about 50° above or below that of the unknown itself, and is not normally recommended. Fractional distillation is preferred but, to be effective, the fractionating column should be at least 20 cm long, and the distillation carried out very slowly, with careful control of the reflux/take-off ratio. Pure liquids should distil over a range of a few degrees. Provided the quantity is sufficient, try firstly to distil a portion of the unknown; stop the distillation if there is any sign of decomposition, and attempt a similar small-scale distillation under vacuum.

(Preparative gas chromatography is by far the most efficient means of purifying stable, volatile, organic compounds, and is the ultimate choice if the equipment is available; but it will not normally be necessary to resort to such refined techniques within the context of a course in qualitative organic analysis.)

1.1 Appearance and Smell

Record the appearance and smell of the unknown (e.g., yellow needles, sweet smell; colourless plates, odourless, etc.).

The colour of the unknown is often most useful in telling you which groups *cannot* be present. Interpret very cautiously:

Yellow, orange, red: nitro and azo compounds, dicarbonyls
Green: nitroso compounds

Try to identify, in the unknown, one of the following smell types (refresh your 'smell memory' with authentic samples if necessary).

Like phenol	Simple monohydric phenols
Like aniline	Simple aromatic amines
Like benzene	Liquid aromatic hydrocarbons
Like ethyl acetate	Many esters, especially aliphatic
Pungent	Lower aldehydes or acids (and their anhydrides and acid halides)
Like pyridine	Simple nitrogen heterocyclics

Many compounds have unique smells, incapable of adequate description.

1.2   Action of Heat

Heat on a nickel spatula or silica crucible lid. Look for:

(a) *Smoky flame*—compounds with high C:H ratios, principally simple aromatic and unsaturated compounds.

(b) *Clear blue flame*—simple saturated aliphatic compounds.

(c) *Coloured flame*—e.g., yellow sodium flame, green copper flame, etc. (associated with (d)).

(d) *Residue*—heat very strongly to ensure that all carbon is burnt off (residue should not be black); if necessary, reheat with a little ammonium nitrate. If a white or coloured residue remains, identify any metal cations from the salts of carboxylic acids or phenols, or from the bisulphite addition compounds of aldehydes, etc. To avoid contamination from the metal spatula, this must be carried out in a silica crucible or lid.

(e) *Burnt sugar smell*—polyhydroxy compounds (principally carbohydrates and derivatives, but also tartrates or citrates).

1.3   Melting Point or Boiling Point

Determine melting point or boiling point. If you have not already purified the substance by distillation or recrystallization, record carefully whether:

(a) substance melts (or boils) over a narrow range, say 2°; substance likely to be pure.

(b) solid does not melt up to 360° (may be a salt).

(c) solid melts first, then decomposes, e.g., becomes dark, or water is given off (useful information in a number of cases). Record this as, e.g.: m.p. 198°d.

(d) solid decomposes before it melts (many complex substances do this, e.g., carbohydrates). Record this as, e.g.: dec. 198°.

1.4 Detection of N, S, P, Halogen—The Sodium Fusion Test

WARNING Some polynitro compounds (e.g., picrates) and polyhalogen compounds (e.g., chloroform) can react explosively with molten sodium. You would normally be warned if your unknown were dangerous, but the following procedure can be applied to such substances—*if carried out as described.* BE SAFETY CONSCIOUS—*wear safety spectacles and carry out the fusion with a safety screen or the fume cupboard front in front of your face.*

Details of the chemistry of the following reactions are discussed in chapter 8, page 166. In the bottom of a small test-tube (*ca.* 60 × 10 mm) place freshly cut sodium metal (rice grain size). Hold the test-tube in a clamp or test-tube holder (*not* in your hand), and melt the sodium over a small flame. Keep the sodium melted, and add the substance:

> *Solids*—20 mg, added in 2–3 mg portions
> *Liquids*—3–4 drops from a m.p. capillary tube.

Heat the mixture strongly for 3 minutes (to burn off unchanged organic matter and to fuse unchanged sodium into the glass) then plunge the red-hot tube into distilled water (10 cm^3) contained in a mortar or large test-tube (2·5 cm diameter); hold a wire gauze between yourself and the tube while doing this.

Boil the resulting solution and filter; use the filtrate ('fusion solution') to test for the following. Note that the filtrate should be clear and colourless—if it is not, then you should discard it and begin afresh, taking care not to add too much substance, and heating the tube strongly for a full 3 minutes after the addition of the substance.

(a) Nitrogen

Add ferrous sulphate (0·1 g) to 1 cm^3 of fusion solution in a test-tube. Boil (1 minute) and acidify with dilute sulphuric acid. A blue or green solution is positive for nitrogen.

$$(\text{N} \longrightarrow \text{CN}^- \longrightarrow \text{Fe(CN)}_6{}^{4-} \longrightarrow \text{Fe}_3[\text{Fe(CN)}_6]_4, \text{blue})$$

(b) Sulphur

(i) *Nitroprusside Test.* Dissolve a minute amount of sodium nitroprusside in cold water (1 cm^3) and add fusion solution (1 cm^3). A purple colour indicates sulphur (see chapter 8, page 166).

(ii) *Lead Sulphide Test.* Acidify fusion solution (1 cm^3) with glacial acetic acid and add lead acetate solution (1 cm^3). A black precipitate of PbS indicates sulphur.

$$(S \longrightarrow S^{2-} \longrightarrow PbS, black)$$

(c) Halogens and Phosphorus

Acidify fusion solution (1 cm^3) with concentrated nitric acid (1 cm^3), and add aqueous silver nitrate (1 cm^3); boil the solution briskly to half bulk to ensure complete removal of HCN and/or H$_2$S (if N and/or S present). A white or yellow precipitate indicates halogen; AgCl is white, AgBr is very pale yellow, and AgI is distinctly yellow.

$$(X \longrightarrow X^- \longrightarrow AgX—white to yellow)$$

Identification of halogen(s) and phosphorus is unambiguously carried out by the following:

(i) **Iodine.** To fusion solution (2 cm^3) add glacial acetic acid (1 cm^3). If N or S is present, boil briskly to half bulk to expel HCN and/or H$_2$S. Add carbon tetrachloride (1 cm^3) and 1 crystal of sodium nitrite; shake the tube and look for a purple colour in the carbon tetrachloride layer, indicating iodine. If iodine is absent, draw off the aqueous layer and proceed to (ii). If iodine is present, add more sodium nitrite (10 mg) to the tube, and extract the aqueous layer with carbon tetrachloride until no more iodine is extracted. Check that all the iodine has been removed by adding another crystal of sodium nitrite, warming, and examining the carbon tetrachloride layer. Draw off the aqueous layer and proceed to (ii).

$$(2I^- \longrightarrow I_2)$$

(ii) **Bromine.** Boil the aqueous layer from (i) until no more nitrous fumes are produced, then cool. Add lead dioxide (0·5 g) and place fluorescein paper (filter paper dipped in ethanolic fluorescein) across the mouth of the tube: heat gently. If the fluorescein turns pink (eosin), bromine is present. Boil the solution until no more bromine is evolved (check with fluorescein paper), then filter off the lead dioxide and proceed to (iii).

$$(2Br^- \longrightarrow Br_2)$$

(iii) **Chlorine.** To the filtrate from (ii) add aqueous silver nitrate (1 cm^3). A white or grey precipitate indicates chlorine.

$$(Cl^- \longrightarrow AgCl)$$

(iv) **Phosphorus.** Boil fusion solution (2 cm^3) with concentrated nitric acid (1 cm^3) for 1 minute; cool, and add aqueous ammonium molybdate. Stand the tube in boiling water for several minutes. A yellow precipitate indicates phosphorus.

$$(P \longrightarrow PO_4{}^{3-} \longrightarrow (NH_4)_3(PMo_{12}O_{40})$$

Mobility of Halogen. If halogen has been shown to be present in the unknown, it is useful to assess its reactivity qualitatively as follows:

Add the unknown to *aqueous silver nitrate:* if a precipitate forms, ensure firstly that it is silver halide (insoluble in nitric acid, soluble in ammonia solution). Halogen compounds which react with, or are ionized by water, react positively (acid halides, quaternary ammonium salts, amine hydrohalides). If no precipitate is observed, then test with *ethanolic silver nitrate*: alkyl halides and aromatic side-chain halides react; iodides rapidly, bromides on warming, and chlorides very slowly, but tertiary, allylic, and benzylic halogens react faster than secondary, which react faster than primary. (See chapter 8, page 167.)

Nuclear aryl halides do not react as above, unless the halogen atom is *ortho* or *para* to activating groups (e.g., nitro).

In tests 1.5, 1.6, and 1.7 which follow, look for *differences* in the solubilities; if possible, perform these tests at the same time to simplify comparisons. These solubility tests enable you to classify the unknown as (a) *acidic*—the substance turns litmus red and is more soluble in alkali than in water (b) *basic*—the substance turns litmus blue *or* is more soluble in acid than in water *or* forms an insoluble salt with acid (see also test 1.9) (c) *neutral*—the substance is neither acidic nor basic as indicated in (a) or (b).

Tests 1.5–1.13 might profitably be carried out at intervals with a series of known compounds: the following are suggested: *p*-toluidine, 2-naphthol, diphenylamine, cinnamic acid, benzamide, benzophenone, catechol (or resorcinol, or hydroquinone).

1.5   Action of Water

Add the unknown to water (1 cm^3). Look for any exothermic reaction (e.g., by hydrolysis of an acid chloride or anhydride). Does the substance dissolve cold? Dissolving substances show Schlieren lines if the tube is observed by strong transmitted light. Test with litmus paper or other pH indicating paper. *Alkaline* reaction given by lower amines (is N present?); also by alkali metal salts of weak acids (is there a residue of metal salt when the unknown is burnt?). *Acid* reaction given by carboxylic acids, sulphonic acids (is S present?), phenols or nitrophenols (is N present?), acid anhydrides or acid halides (is halogen present?), hydrohalides or sulphates of amines (are N and halogen, or N and S, etc., present?), imides. *Acidic impurities are often misleading in this test*, (for instance acidic material in aliphatic aldehydes): do not rely solely on the litmus test for indications of acidic character.

Heat the mixture and cool. If the substance is sparingly soluble in cold water, does it dissolve in hot water? Does it crystallize out unchanged on cooling? Any material crystallizing out with different appearance from the original compound should have its m.p., etc., checked. If different from the original, treat as a separate unknown compound.

1.6   Action of Dilute Acid

Add the unknown to dilute hydrochloric acid (1 cm^3).

Is the unknown *more* soluble than in water (cold or hot)? Most amines are more soluble, but this is not true of all diaryl and triaryl amines (which are only very weakly basic.) Look for Schlieren lines as an indication of solubility.

Smell the solution—try to identify any gas evolved (e.g., SO_2 from bisulphite addition compounds).

1.7   Action of Dilute Alkali

Add the unknown to dilute sodium hydroxide solution (1 cm^3).

Is the unknown *more* soluble than in water (cold or hot)? Acidic substances are more soluble (carboxylic acids, sulphonic acids—is S present? —phenols and derivatives). Look for Schlieren lines.

Smell the solution cautiously, both cold and hot. Is ammonia evolved? NH_4^+ salts give off ammonia with cold or warm alkali; amides or imides from hot alkali.

Does the solution turn dark or gummy when heated? Polyhydric phenols are oxidized by air to dark brown products, aliphatic aldehydes polymerize to yellow-red resins: nitrophenols give intensely yellow solutions (is N present?), and the colour intensity is markedly reduced on acidification of the alkaline solution. (See chapter 6.)

1.8   Action of Bicarbonate

Add the unknown to *cold* sodium bicarbonate solution (1 cm^3).

Is CO_2 evolved? Observe carefully the surface of any undissolved crystals of the unknown—evolution of CO_2 may be very slow. Carboxylic acids (except the higher members), sulphonic acids, and many nitrophenols give this test, and thus can be differentiated from the other acidic substances indicated in tests 1.5 and 1.7.

1.9   Action of Concentrated Sulphuric Acid

Add the unknown to concentrated sulphuric acid (1 cm^3). Use a glass rod to disperse solids in the acid.

Look for: *a white precipitate*: amines form salts, even weakly basic amines like triarylamines (is N present?): *charring*: many polyhydroxy compounds char in cold or warm concentrated sulphuric acid, e.g., carbohydrates, citrates, or tartrates: *strong colours*; simply note this, and retain as possible confirmatory evidence (too many classes give colours for this result to be very useful).

1.10   Action of Bromine—Test for Unsaturation or Easy Substitution

Dissolve the unknown in water (1 cm^3) or chloroform, or carbon tetrachloride. *From a dropper* slowly add a solution of bromine in the same solvent. *Simultaneously run a blank test with solvent alone.*

Decolourization of the bromine (compared with the blank) indicates addition of bromine to a double or triple bond, or substitution in an activated aromatic nucleus (e.g., phenol or amine).

A white precipitate may be the bromo derivative of a phenol or amine (is N present?).

1.11   Action of Permanganate—Test for Unsaturation or Easy Oxidation

Dissolve the unknown in water, or ethanol, or acetone (1 cm^3). From a dropper add 1–2 drops of dilute potassium permanganate. *Simultaneously run a blank test with solvent alone.*

Note if the colour of MnO_4^- is discharged; unsaturated compounds *and* other easily oxidized groups are indicated (e.g., many amines and phenols, dihydric phenols rapidly, α-hydroxy ketones and α-diketones, aldehydes and many of their derivatives).

1.12   Action of 2,4-Dinitrophenylhydrazine—Test for Aldehydes and Ketones

To methanol (3 cm^3) in a test-tube add 2,4-dinitrophenylhydrazine (0·2 g), followed *cautiously* by concentrated sulphuric acid (0·5 cm^3). Shake and warm until the reagent dissolves, then add the unknown (0·2 g), and warm gently. If no precipitate forms immediately, warm for 1 minute, then cool under cold water. If no precipitate forms, add 3–4 drops of water and stand the tube in cold water for 5 minutes. A yellow, orange, or red precipitate indicates aldehyde or ketone, together with some of their derivatives. Amines commonly interfere. (See chapter 8, page 168.)

Differentiation of Aldehyde from Ketone. *Note*: Infrared and n.m.r. evidence is definitive in this: for aldehydes, C—H *str* is near 2800 cm^{-1} (see page 12) and the proton chemical shift position is near 10 δ (see page 128, (f)). Perform the following three tests; typical aliphatic aldehydes give all three positive, but note that some aromatic aldehydes give negative Schiff's or Fehling's tests; that some ketones (especially α-hydroxy ketones and α-diketones) give positive Tollens' test; and that other functions present (e.g., amine) may give positive reactions too. For more details see chapter 8, page 169.

Schiff's test. Shake the unknown with Schiff's reagent (1 cm^3); if necessary, dissolve the unknown in ethanol. Do not heat. Aliphatic aldehydes restore the magenta colour within a few minutes.

Tollens' test. To aqueous silver nitrate (1 cm^3) add dilute sodium hydroxide (1 drop) and then enough dilute ammonia solution to dissolve the precipitate of silver oxide thus formed. Add the unknown, and stand the tube with occasional shaking in a beaker of boiling water. Do not boil directly over a flame. The formation of a silver mirror on the test-tube walls is positive.

CAUTION: *after completing the test, wash the mixture down the drain thoroughly, and remove any silver mirror with dilute nitric acid.* Residues from this test may contain silver fulminate, which is explosive.

Fehling's test. Mix Fehling's solutions A and B (1 cm^3 of each), add the unknown, and boil gently. Aldehydes *and other easily oxidized groups* reduce the blue copper(II) solution to green copper(I) solutions or to a precipitate of red copper(I) oxide.

1.13   Action of Soda-Lime

In a hard glass test-tube mix the unknown with powdered soda-lime; cover the mixture with a layer of soda-lime and heat strongly. Observe and smell. Concentrate on odours *different* from that of the unknown. Look particularly for the evolution of:

Ammonia (check with litmus or other indicator paper): evolved from NH$_4^+$ salts, amides, or imides.

Benzene (or other aromatic hydrocarbon): soda-lime causes decarboxylation of carboxylic acids—the odour of the residual molecule after decarboxylation will be detected.

It is often difficult to identify the smell of products from the soda-lime test; give weight only to the production of smells which you recognize.

2

Classification of an Unknown Compound—Chemical and Infrared Evidence

Deduce, first, the nature of the carbon skeleton of each unknown (section 1).

The unknown is then allocated initially to one of the classes 2–11 below; *this depends on the accurate performance of the Lassaigne's sodium fusion test.*

Thereafter, subclassification into functional groups is carried out on the basis of solubility, basicity, etc., taken together with relevant features of the infrared spectrum.

1.	The carbon skeleton			page	10
2.	Unknown contains CH only			,,	11
3.	,,	,,	CHN	,,	19
4.	,,	,,	CH Halogen	,,	29
5.	,,	,,	CHS	,,	30
6.	,,	,,	CHN Halogen	,,	31
7.	,,	,,	CHNS	,,	33
8.	,,	,,	CHS Halogen	,,	35
9.	,,	,,	CHNS Halogen	,,	36
10.	,,	,,	CHP	,,	37
11.	,,	,,	a Metal	,,	37

Oxygen may or may not be present in all the compounds classified as above.

The numbers given in parentheses below refer to the tests given in chapter 1 (e.g., 1.12 means chapter 1, test 12).

For the infrared evidence, the abbreviations used are those commonly adopted, namely: *str*—stretch, *def*—deformation, *s*—strong intensity absorption, similarly *m*, *w*, and *v* are medium, weak, and variable respectively. Unless expressly stated otherwise, *comments on infrared absorptions refer to survey spectra run as thin films (liquids) or as KBr discs, or mulls for solids.*

2.1 The Carbon Skeleton

For all classes, decide whether the skeleton of the compound is (a) aromatic or (b) aliphatic. Aromatic compounds may have aliphatic side-chains, etc. Detect also (c) alkenes and alkyne groups.

(a) **Aromatic Compounds.** Burn with a smoky flame usually. Aromatic compounds *always* show two groups of bands in the infrared spectrum; the first group is associated with the in-plane stretching vibrations of the nuclear bonds, and occurs in the region 1470–1640 cm^{-1} (three bands are most frequently observed, as shown on the charts; a fourth band may also be seen around 1450 cm^{-1}, but quite often only two bands can be clearly discerned). The second group arises from the out-of-plane deformations (bending) of the nuclear C—H bonds, and these absorptions lie between 690–860 cm^{-1}. Use the correlation charts to ascertain if these bonds are present; if not, the compound has a wholly aliphatic carbon skeleton. If the compound is aromatic, reach a tentative conclusion about the substitution pattern of the aromatic nucleus, using initially the bands 690–860 cm^{-1}, which apply *only* to benzenoid compounds. Do not apply these correlations to heterocyclic aromatic systems: see note below.

Two other regions should also be scanned for aromatic absorptions (i) the C—H *str* absorption produces weak bands just above 3000 cm^{-1}, characteristic of nuclear C—H bonds: this absorption may appear as a shoulder on other stronger absorptions, or may be entirely swamped by them: (ii) weak bands also appear between 1600 and 200 cm^{-1} (overtone and combination bands), but these are frequently unobserved unless a large sample is used. The relative intensity of these bands is evidence of the substitution in the ring: taken in conjunction with the bands at 690–860 cm^{-1}, it is often possible to deduce the substitution pattern at this stage. (Many exceptions to these correlations exist, and such conclusions may require revision in the light of subsequent evidence.) A sample set of spectra should be recorded on the same instrument to show the overtone and combination bands in known compounds: for simplicity use liquid samples, which can more easily produce strong sharp spectra for reference purposes.

Note: N.m.r. evidence is more reliable for deducing ring substitution patterns. (See pages 130–132.)

Heterocyclic Aromatic Systems show C—H *str* and C—C *str* absorptions within the ranges shown on the charts for benzenoid compounds: C=N *str* occurs also within the quoted C—C *str* ranges.

In addition, pyridine-based systems show C—H *def* absorptions as follows: near 1200 cm^{-1}(s) and 1000–1100 cm^{-1}(s). Deduction of the substitution patterns from the position of other C—H *def* absorptions is not recommended.

(b) **Aliphatic Compounds.** Saturated aliphatic compounds burn with a pale blue smokeless flame usually.

Identify on the infrared spectrum aliphatic C—H *str* (strong bands below 3000 cm^{-1}), and the C—H bending vibrations around 1400 cm^{-1}, (see correlation charts). Try to infer the presence of particular alkyl groupings (e.g., isopropyl, t-butyl): interpret cautiously at first, but when a final identification of the unknown compound has been made, you should re-examine the spectrum, and make more definite assignments of the aliphatic vibrations. Many skeletal vibrations occur in the range 800–1200 cm^{-1}, but these are of little diagnostic value and are not shown on the charts.

Aromatic Compounds with Aliphatic Side-Chains. Burn with a smoky flame and show infrared features characteristic of classes (a) and (b).

(c) **Alkene and Alkyne Groups.** Decolourize Br$_2$ and MnO$_4^-$ (1.10 and 1.11). If the compound is aromatic, it will be difficult to detect an alkene group from the infrared evidence, since many absorptions are very similar in the two classes. Use the correlation charts to search for alkene C—H and C=C *str* bands: if the compound is entirely aliphatic these should be easily distinguished. (For n.m.r. evidence see page 126.)

For alkynes, look for the sharp acetylenic C—H *str* band around 3300 cm^{-1} (for terminal alkynes only) and for the C≡C *str*: the position and intensity of the C≡C *str* changes from mono to disubstituted alkynes as indicated on the charts. (For n.m.r. evidence see page 126.)

2.2   Compounds Containing C, H, and possibly O

NEUTRAL COMPOUNDS

(2a) Aldehydes and Ketones (and Quinones)

(2b) Esters (and Lactones)

(2c) Carbohydrates

(2d) Alcohols

(2e) Ethers and Hydrocarbons

ACIDIC COMPOUNDS

(2f) Carboxylic Acids

(2g) Carboxylic Anhydrides

(2h) Phenols and Enols

Neutral Compounds

Use the correlation charts to search the infrared spectrum for the strong C=O *str* band (1650–1780 cm^{-1}). If present, the compound belongs to one of the classes (2a) or (2b).

If C=O *str* absorption is absent, search the spectrum for O—H *str*: if present, the compound belongs to class (2c) or (2d).

If neither C=O *str* nor O—H *str* is present, the compound belongs to class (2e).

Further confirmation of these classifications is given below.

Long-chain fatty acids (C$_{10}$ upward) have very weak acidic properties and may mistakenly be classified as neutral: infrared evidence for carboxyl is definite enough to enable these to be classified correctly (2f).

(2a) **Aldehydes and Ketones, including Quinones.** These will be detected chemically by the 2,4-D.N.P. test (1.12), and aldehydes will probably be distinguished by Tollens', Fehling's, and Schiff's tests (1.12a). Lower aliphatic aldehydes are water soluble and have pungent smells: they are virtually always contaminated with the corresponding acid, and therefore CO_2 is liberated from bicarbonate solution. Note that (non-aqueous) formaldehyde is a gas, and acetaldehyde has a b.p. of 20°. Aromatic aldehydes with the CHO group on the nucleus do not generally reduce Fehling's solution (1.12a). Lower aliphatic ketones are water soluble and have sweet smells. α-Hydroxy ketones behave normally except for their ease of oxidation; they reduce Tollens' and Fehling's reagents (1.12a).

The chemical behaviour of dicarbonyl compounds varies considerably depending on the relative positions of the two groups: α-dicarbonyls are usually pale yellow but otherwise behave normally (see, however, chapter 8 page 169): β-dicarbonyls exhibit keto-enol tautomerism (see infrared evidence below): γ-dicarbonyls etc., behave normally.

Quinones are yellow or red solids, but commercial samples may be discoloured. Few general chemical reactions are available, but spectroscopically they are similar to normal α,β-unsaturated ketones.

The most significant difference between the spectra of aldehydes and ketones is the C—H *str* of the CHO group around 2800 cm^{-1} (see charts), but this varies considerably in intensity. Often two bands can be seen. N.m.r. evidence for aldehydes is definitive. See page 128, (f).

Use the evidence about the carbon skeleton obtained in 2.1 to predict the expected range within which the C=O *str* absorption should occur; the precise position of the band will depend on the nature of the carbon residues attached to the carbonyl group (see charts for average values).

Note that chelation produces a marked lowering of the C=O *str* absorption (e.g., in *o*-hydroxybenzaldehydes, etc.). If a ketone or aldehyde is detected chemically, and no absorption is found in the range 1650–1780 cm^{-1}, look for strong absorption:

ca. 1540 cm^{-1}, C=O *str* of β-diketones. These will be to some extent enolic, and should be examined accordingly (2h).

ca. 3300 cm^{-1}, O—H *str*: possibly a carbohydrate (2c).

Iodoform Test. For ketones, test for the presence of the CH_3CO group as follows: dissolve or suspend the unknown (0·2 g) in 10 per cent potassium iodide solution (3 cm^3), add dilute sodium hydroxide (0·5 cm^3), and then strong sodium hypochlorite solution (2 cm^3). Shake the solution, and if necessary warm for a few minutes. Iodoform, CHI_3, will precipitate as yellow crystals or a yellow turbidity if CH_3CO is present. Why do acetaldehyde and propan-2-ol give a positive test? It is important to know why.

(2b) Esters and Lactones. Chemical confirmation is not always easy, while infrared evidence is definitive.

Most simple esters of aliphatic acids have pleasant, fruit-like odours; most aromatic esters have less pleasant odours. Most of the liquid esters are sparingly soluble in water, but simple formate esters hydrolyse rapidly in water to the alcohol and formic acid. The presence of other functions will modify these points.

Hydroxamic Acid Test. This is the only reasonably specific chemical test for an ester, but some esters give a negative result, while positive results are obtained from some non-esters, including amides and anhydrides of carboxylic acids. In this test, the ester is converted to a hydroxamic acid ($R.CO.NHOH$), which gives coloured complexes with $Fe(III)$ ions.

To dilute sodium hydroxide solution (5 cm^3), add the unknown (*ca.* 10 mg) and hydroxyammonium chloride (hydroxylamine hydrochloride) (0·1 g); boil the mixture for 2 minutes, then acidify with dilute acetic acid and add a few drops of aqueous ferric chloride. If a hydroxamic acid has been formed, a red or violet colour will develop. (A blank test should be carried out simultaneously). The reaction involves base-catalysed attack of hydroxylamine on the ester. See chapter 8, page 169.

This test must only be used to detect an ester group in the absence of anhydride, acid halide, amide, or nitrile; the presence of these groups in the compound will, however, be easily inferred, and the unknown will be classified under the appropriate class.

Infrared absorptions of esters are very precisely known, and such evidence is almost always diagnostic. Both the $C=O$ and $C—O$ *str* absorptions vary with the particular groups attached to the ester function: use the deductions about carbon skeleton (see 2.1) to assist in the inspection of the correlation charts for ester absorptions. (See Chart 2b.)

Ketones are occasionally wrongly classified as esters if the 2,4-D.N.P. test has been incorrectly carried out. If there is any doubt about identifying the compound as an ester, repeat the 2,4-D.N.P. test exactly as prescribed in 1.12.

(2c) Carbohydrates. When burnt, simple carbohydrates produce the characteristic smell of burnt sugar. All are colourless solids, or syrupy liquids. Monosaccharides and disaccharides are insoluble in ether and very soluble in water. Some polysaccharides give gummy solutions in water; starch (a mixture of polysaccharides) gives a cloudy suspension, while cellulose is insoluble. Most will give a positive 2,4-D.N.P. test, and will reduce Tollens' and Fehling's solutions under the conditions given in 1.12. The ease with which this occurs can be a pointer to the structure of certain disaccharides. The absence of C=O *str* in the infrared is significant.

Molisch's Test. This is the classical confirmatory test for carbohydrates: dissolve a few mg of 1-naphthol in ethanol ($0.5 \, cm^3$) and add a few drops of this solution to a solution of the unknown (20 mg) in water ($2 \, cm^3$). Shake and carefully pour concentrated sulphuric acid down the side of the tube to form a lower layer; a red or violet colour is formed at the interface on standing, and on shaking the tube the mixture becomes blue-violet. See also chapter 8, page 170.

Normally there is no difficulty in classifying an unknown as a carbohydrate, and little additional information comes from the infrared except:

O—H *str*—Very strong, broad absorption from 3200–3400 cm^{-1}

C—H *str*—Very strong absorption around 2900 cm^{-1}, clearly separated from the O—H *str*: many strong bands around 1000–1100 cm^{-1}, associated with C—O *str* and O—H *def*. For polysaccharides, etc., a β-glycosidic link gives rise to absorption around 890 cm^{-1} (*m*): α-glycosides absorb around 844 cm^{-1} (*m*).

Polysaccharides are best identified from certain of their chemical properties. For others, optical rotation is a valuable aid, and this is discussed along with the preparation of derivatives.

Derivatives: page 61
Table: *Carbohydrates*, 4.8

(2d) **Alcohols.** The lower alcohols only (up to C_3) are completely soluble in water: higher alcohols are progressively less soluble, and the solid alcohols are virtually insoluble.

Dihydric alcohols are solids or viscous liquids (because of hydrogen bonding): they are water soluble, and almost insoluble in non-polar solvents (ether, hexane, etc.).

Chemical confirmation of an alcohol is often difficult in practice, and the best evidence comes from infrared or n.m.r. studies, but carry out the following three tests and **be prepared to interpret the observations cautiously**.

Action of Acetyl Chloride. If the unknown is a liquid, dry a little by shaking with anhydrous magnesium sulphate for a few minutes: if a solid, dissolve a little in chloroform ($0.5 \, cm^3$): now add three or four drops of acetyl chloride. Look for a vigorous evolution of hydrogen chloride:

ROH + CH$_3$COCl $\longrightarrow$ CH$_3$COOR + HCl. Finally add 1 ml of water and try to identify the sweet smell of any ester (not all esters have pleasant smells).

Tertiary alcohols react differently, giving acetic acid and the tertiary halide. See chapter 8, page 170.

Lucas' Test. The rate of conversion of an alcohol to an alkyl halide follows the sequence tertiary > secondary > primary. (See chapter 8, page 170.) Lucas' reagent may be supplied; if not, proceed as follows.

To concentrated hydrochloric acid (5 cm^3) add anhydrous zinc chloride (2 g) and boil for 1 minute: cool to room temperature.

Add the alcohol (1 cm^3) and shake the tube vigorously. Stand the tube in a water bath at 27–28°, and watch for the production of alkyl halide (mixture becomes cloudy).

Tertiary alcohols react almost immediately: secondary alcohols (and benzylic or allylic alcohols) take up to 5 minutes: primary alcohols do not usually react in less than 20 minutes.

Many alcohols behave anomalously, e.g., isopropyl alcohol, dihydric alcohols, and higher alcohols.

Iodoform Test. Carry out the test as performed on ketones (page 12); a positive iodoform reaction indicates the group CH$_3$—CHOH (easily oxidized to CH$_3$CO).

Infrared evidence will confirm conclusively the presence of OH: the substance *must* be dry. (For n.m.r. evidence see page 128, (g).)

Liquid samples (liquid films on rock salt flats) or KBr discs of solid alcohols will normally show a moderately broad single absorption around 3300 cm^{-1}. This arises from the hydrogen bonded O—H *str* vibration. Stronger hydrogen bonding will lower the frequency and broaden the band, but this is most commonly encountered in chelated compounds, which may have carbonyl functions also present in the molecule. To observe the free O—H *str* of alcohols (around 3600 cm^{-1}) the sample must be examined in solution (e.g., 5 per cent in carbon tetrachloride). (Solution spectra are not given detailed coverage in this book.)

The pattern of O—H *def* absorptions (around 1200 cm^{-1}) may help to confirm whether the alcohol is primary, secondary, or tertiary (see charts, 2d). Interpret these patterns with reservation unless other evidence supports your deductions.

Derivatives: page 62
Tables: *Aliphatic Alcohols,* 4.9
              Aromatic Alcohols, 4.10

(2e) **Ethers and Hydrocarbons.** If a neutral compound contains only C, H, and possibly O, and cannot be classified into any of the groups 2a–2d, then it is either an ether or a hydrocarbon. Epoxides are treated as ethers.

A general chemical method of differentiating these two classes is not possible, since many of their properties overlap, but infrared evidence can often be definitive.

All simple ethers and hydrocarbons are substantially insoluble in water; all ethers dissolve in concentrated sulphuric acid to form oxonium salts (unsaturated hydrocarbons and many polynuclear hydrocarbons dissolve too, the latter often giving coloured radical cations in solution). Chemical tests will detect any unsaturation (decolourization of Br_2 and MnO_4^-). Further distinctions must rest with infrared.

Hydrocarbons have relatively simple infrared spectra, while ethers have strong C—O *str* absorptions (see charts). For hydrocarbons, the infrared spectrum should be examined using the chart for 'Carbon Skeleton' alone. Confirm your assessment of the carbon skeleton.

Examine the ultraviolet absorption spectrum of any solid aromatic hydrocarbon (which may be polynuclear) and compare it peak by peak with the data given on table 6.3. Simple alkyl substituents on a hydrocarbon molecule make little change in the spectrum: this is amplified in chapter 6.

> *Derivatives:*   page 64
> *Tables:*   *Aliphatic Ethers*, 4.11
> *Aromatic Ethers*, 4.12
> *Alkanes and Cycloalkanes*, 4.13
> *Alkenes and Cycloalkenes*, 4.14
> *Alkynes*, 4.15
> *Aromatic Hydrocarbons*, 4.16

Acidic Compounds

The ferric chloride test (see below) is usually sufficient to distinguish phenols and enols from other acidic substances; phenols and enols give violet to green colours. Lower aliphatic acids may give reddish colours, but only in neutral solution.

Also, carboxylic acids and their anhydrides are usually sufficiently acidic to liberate CO_2 from aqueous bicarbonate (but see below).

For all classes, a number of distinguishing features in the infrared spectrum makes differentiation relatively straightforward.

Keto acids are investigated as for simple carboxylic acids. Most behave normally as ketones, and hence give a positive 2,4-D.N.P. test: β-keto acids decarboxylate on heating to *ca.* 150° and this may be observed as frothing during the determination of m.p. Esters of phenolic acids are identified initially as phenols.

(2f), (2g) **Carboxylic Acids and Carboxylic Acid Anhydrides.** Liquid acids are either pungent (like acetic acid) or foul smelling (like butyric acid). Most liquid acids are soluble in water, and all dissolve in dilute sodium hydroxide and liberate CO_2 from dilute sodium bicarbonate solution.

Solid acids range in water-solubility from very soluble (like citric acid) to insoluble. All are soluble in dilute NaOH, and all but the long-chain fatty acids effervesce with sodium bicarbonate: *both of these reactions may occur very slowly.*

Anhydrides behave in aqueous tests as acidic, because of hydrolysis to the corresponding acid; indeed, most samples of anhydride are contaminated with the acid. Thus they turn litmus red, dissolve in alkali, and effervesce with bicarbonate.

The most satisfactory chemical method of detecting an anhydride group is to isolate the derived acid: warm the unknown with a small amount of dilute alkali until it dissolves. Make just acid with hydrochloric acid, and if any solid is precipitated, isolate it and show it to be a carboxylic acid (infrared) and different from the original unknown. If there is no precipitation, extract with ether (2 cm^3) evaporate off the ether, and compare the residue with the original as above.

Note: some simple formate esters are so easily hydrolysed that they may appear to behave as acids in those tests involving aqueous reagents. Any liquid having a fruit-essence smell, but showing the reactions of a carboxylic acid, is likely to be an ester of formic acid: examine it spectroscopically as shown under Esters (2b).

Carboxylic Acids show very broad O—H *str* absorption, because of hydrogen-bonded dimeric association: the series of peaks may stretch from 3000 cm^{-1} to 2500 cm^{-1}, with considerable fine structure superimposed on them. Absorption of this type occurring *with* a carbonyl peak is very characteristic of carboxylic acids, and is sufficient to identify them. N.m.r. evidence is also characteristic: see page 128, (o).

Highly enolic β-dicarbonyl compounds have similar infrared spectra, but are distinguished by other features (see below).

The position of the carbonyl absorption should be checked against your evidence for the carbon skeleton of the compound, since the frequency of C=O *str* is markedly affected by such factors as conjugation (see charts, 2f).

Phenolic Acids show the spectroscopic and chemical properties of carboxylic acids and phenols; in the special case of *o*-hydroxy carboxylic acids (e.g., salicylic acid) the frequency of the C=O *str* is lowered by chelation as shown on the charts.

Carboxylic Acid Anhydrides should show no O—H *str*, but the sample may well be contaminated with some of the free acid. The most characteristic feature is the presence of *two* bands in the C=O *str* region: these should be related in position to the nature of the carbon skeleton (see correlation charts). The two C=O *str* bands probably arise through coupling of the vibrations of the two carbonyl groups; C—O *str* also varies from cyclic to acyclic anhydrides as shown on the charts.

Derivatives: pages 68–69
Tables: *Aliphatic Carboxylic Acids*, 4.17
             Aromatic Carboxylic Acids, 4.18
             Carboxylic Acid Anhydrides, 4.19

(2h) **Phenols and Enols.** Phenols, if simple, have the characteristic odour of carbolic acid (i.e., phenol), and are liquids or low melting solids. Phenols of this type are vesicant, and should be handled with care. Higher phenols and dihydric phenols may have little smell.

The OH group activates the aromatic nucleus toward electrophilic reagents, and all phenols react rapidly with bromine-water, etc., giving nuclear substitution. Dihydric phenols are particularly reactive, and are oxidized by MnO_4^- and even by Ag^+, as in Tollens' reagent.

Enols will easily be detected if the position of the keto-enol equilibrium is not too far towards the keto form, but most show distinct carbonyl activity (see below). For n.m.r. evidence see page 129, (p).

Phenols and enols dissolve in dilute sodium hydroxide, but do not liberate CO_2 from bicarbonate. For n.m.r. evidence see page 128, (i).

Ferric Chloride Test. Dissolve or suspend the unknown in water or aqueous methanol and add 2 drops of ferric chloride solution. Phenols and enols give colours ranging from green through red to purple. Dihydric phenols may be oxidized, giving intensely coloured or black products.

In the infrared study of phenols, most weight attaches to the detection of O—H *str* absorption. (*The compound must be aromatic.*)

Simple phenols and dihydric phenols show the broad absorption of bonded O—H *str* around 3300 cm^{-1}; free O—H *str* is only shown in dilute solution (3600 cm^{-1}).

Phenolic Acids, are detected and identified as carboxylic acids, the only phenolic chemical properties detected are the Fe^{3+} colours. When heated with soda-lime they are decarboxylated to phenols which may be detectable by smell. The phenolic O—H *str* absorption will usually be swamped by carboxyl O—H *str*, and will only be seen in dilute solution spectra, when it moves to 3600 cm^{-1}. Chelated phenolic acids (e.g., salicylic) do not show free O—H *str* even in dilute solution.

Enols may show absorptions in the infrared due to the associated keto forms. In compounds which are highly enolic (e.g., $PhCOCH_2CHO$ or even $CH_3COCH_2COCH_3$) the O—H *str* absorption may be as broad as in carboxylic acids because of the intramolecular hydrogen bonding, but in these cases the C=O *str* will be low (1540–1650 cm^{-1}) compared to carboxyl C=O *str*.

Less highly enolic compounds show the chemical and spectroscopic properties of keto and enol forms, so that ethyl acetoacetate shows normal keto and ester C=O *str* (1720 and 1740 cm^{-1}) together with the chelated ester C—O *str* of the enol form (1650 cm^{-1}).

Examine all enolic compounds for normal aldehyde, ketone, and ester absorptions: try to get some idea of the equilibrium position from the intensity of the infrared absorptions.

Examine them chemically for aldehyde, ketone, and ester behaviour, and see also chapter 8, page 170.

Derivatives:	page 70
Tables:	*Phenols*, 4.20
	Aliphatic Enols, 4.21
	Aromatic Enols, 4.22

2.3   Compounds containing C, H, N, and possibly O

This is the largest single group of organic functions, and very careful sub-division is necessary. *Solubility distinctions alone do not produce sharp divisions among the various classes, and the student must interpret his observations cautiously.*

Should the result of any test be ambiguous, be prepared to look for a compound in more than one sub-class until clear confirmation of identity is obtained.

Three principal groups can be identified on the basis of solubility in water (1.5), dilute acid (1.6), and dilute alkali (1.7):

Basic Compounds	page 20
Neutral and Amphoteric Compounds	page 24
Acidic Compounds	page 28

If the compound is coloured, even though very pale yellow, read the notes below on *Nitro compounds, Azo compounds,* and *Nitroso compounds.*
Nitro Compounds: where a nitro group is present in a molecule *with* one of the other nitrogen containing functions of class 3, the compound will automatically be allotted to that class (e.g., nitro amides will be detected as amides, etc.). In general, the nitro group does not interfere with the detection of the other function *except* for nitro amines, which are much less basic than simple amines.

The compound may be a nitro derivative of a class 2 compound (e.g., nitro hydrocarbons or nitro carboxylic acids), and therefore any neutral or acidic compound in which you are unable to detect a class 3 functional group should be examined systematically for a class 2 function. These nitro compounds can be detected by looking for the NO_2 absorptions in the infrared spectrum; other evidence is given below for specific neutral and acidic nitro compounds.
Azo Compounds: these are always strongly coloured. Any compound which is yellow, orange, or red, may be an azo compound (see 3p).

Nitroso Compounds: the commonest nitroso compounds are nitroso substituted phenols (yellow to brown) and nitroso substituted aromatic tertiary amines (deep green). (See 3q and 4.3q.)

Basic Compounds

(3a) Primary Aliphatic Amines

(3b) Secondary Aliphatic Amines (including Heterocyclics)

(3c) Tertiary Aliphatic Amines (including Heterocyclics)

(3d) Aromatic Side-Chain Primary Amines

(3e) Primary Aryl Amines and Diamines

(3f) Secondary Aromatic Amines

(3g) Tertiary Aromatic Amines (including Aromatic Heterocyclics)

(3h) Hydrazines and Semicarbazides

(3i) Imines (including Schiff's Bases) and Aldehyde-Ammonias.

This large group includes compounds which are clearly basic by virtue of an alkaline reaction in aqueous solution, *or* show marked solubility in dilute acid compared with water, *or* show obvious salt formation with dilute or concentrated acid.

With so many functional groups involved, the chemical and spectroscopic examination must be carried out carefully, and with the realization that the properties of some of these groups overlap to a considerable extent.

The following scheme sets out in a simplified way the process of subdividing classes (3a)–(3i). The n.m.r. spectrum will confirm the number of N—H protons present (page 128).

Basic C, H, N Compounds

1. *Nitrous acid reaction:* Aryl NH_2 gives diazo coupling

    Other primary NH_2 evolve N_2

    Secondary NH give N-nitroso derivatives

    Tertiary aromatic amines give *p*-nitroso derivatives

2. *Infrared spectrum:* Aliphatic or Aromatic?

    Primary NH_2—2 or 3 N—H *str* bands

    Secondary NH—1 N—H *str* band

    Tertiary—no N—H *str*

3. *Strong reducing properties:* Hydrazines

    Semicarbazides

    Diamines

4. *Positive 2,4-D.N.P. test:* Imines (including Schiff's Bases)

    Aldehyde-ammonias

General. Lower aliphatic amines smell like ammonia or (especially if tertiary) like fish. Most aromatic amines give rapid decolourization of bromine-water, because of ring substitution; they also react with MnO_4^-

because of easy oxidation to quinones. Many reduce Tollens' reagent for the same reason. Hydrazines and semicarbazides are strong reducing agents. Class 3i compounds give positive 2,4-D.N.P. tests.

On all basic compounds carry out the test with nitrous acid as an aid to identification of the functional group:

Action of Nitrous Acid. Dissolve or suspend the compound (0·1 g or 0·1 cm^3) in hydrochloric acid (2 cm^3 concentrated acid plus 3 cm^3 water); warm if necessary to aid dissolution, then cool in ice to around 5°. Slowly add 10 per cent aqueous NaNO$_2$ solution until an excess of nitrite is present (starch–iodide test). Keep the temperature below 10° during the addition of nitrite.

One of the following will be observed: (the chemistry of these reactions must be understood fully to derive maximum value from the observations; see chapter 8, page 170):

(i) *Nitrogen evolved:* this should be continuous and vigorous, and the solution should remain clear. Do not confuse with the very slow evolution of nitrous fumes from the nitrous acid. Most primary aliphatic amines and side-chain aromatic primary amines react thus (3a and 3d).

(ii) *No apparent reaction:* solution remains clear. Ensure no large excess of nitrite is present. Test carefully with starch–iodide paper; excess nitrite can be removed by adding urea until no further effervescence occurs. Now add 2 cm^3 of the solution to a solution of 2-naphthol (50 mg) in dilute NaOH (2 cm^3). If an intense orange or red dye is produced, the compound is a primary aryl amine (3e). If the original amine is simply regenerated on pouring into alkali, it is probably tertiary (3c or 3g) but may be secondary (3b or 3f). A few compounds give red colours in this test, but not the intense red of a positive test as given by primary aryl amines: if in doubt, repeat the test with aniline for comparison.

(iii) *Solution becomes turbid or deep red:* a yellow oil or solid may be the N-nitroso derivative of a secondary amine (3b or 3f), or compounds of similar appearance from *o*-diamines (3e). A dark brown solid is probably the self-condensation product from a *m*-diamine (3e). If the solution becomes very deep red, with or without any precipitation, add alkali: tertiary amines (3g) give *p*-nitroso compounds which are intense deep green. Their salts, in acid solution, are deep red. (See 4.3g.)

The Infrared Spectrum. Make certain the compound is dry, since water absorbs broadly around 3500 cm^{-1}.

The primary NH$_2$ group is very easily detected spectroscopically; use the correlation charts to examine the spectrum for *two* N—H *str* bands (hydrogen bonding may give rise to a third), and the strong N—H *def* absorption (3a, 3d, 3e). Aryl NH$_2$ groups also show strong C—N *str*. N—H *str* bands are sharper than O—H *str* bands and are, because of weaker hydrogen bonding, less affected by solvent changes than O—H bands. For liquid films and KBr discs, the observed bands are the hydrogen bonded absorptions.

Secondary amines show only *one* N—H *str* band; the N—H *def* absorption is usually weak and of little diagnostic value. Aryl amines show C—N *str*.

Tertiary amines show no N—H *str* absorption, and cannot easily be identified by infrared spectroscopy. Aryl amines show C—N *str*.

Semicarbazides show C=O *str* (3h), but this may be difficult to distinguish from N—H *def* around 1650 cm^{-1}. Otherwise, hydrazines and semicarbazides have similar spectral features to amines: i.e., if NH_2 is present, two N—H *str* bands, etc.

Nitro compounds show characteristic absorptions, as shown on the charts.

(3a) **Primary Aliphatic Amines.** Lower members are water soluble and smell like ammonia; methylamine and ethylamine are gases. They dissolve readily in dilute acids to form salts, and react with nitrous acid to give diazo compounds which immediately lose nitrogen. Confirmation from infrared evidence is straightforward (two or three N—H *str* bands).

(3b) **Secondary Aliphatic Amines, including Heterocyclics.** Similar in most respects to the 3a compounds, they can usually be differentiated by infrared evidence alone (only one N—H *str* band).

(3c) **Tertiary Aliphatic Amines.** The most notable feature of these is the absence of N—H *str* or N—H *def* in the infrared, provided the sample is relatively free of other amine contaminants which do show these features. Otherwise, they have similar physical properties to the (3a) and (3b) compounds.

> *Derivatives:*   page 71
> *Tables:*   *Primary Aliphatic Amines,* 4.23
>   *Secondary Aliphatic Amines,* 4.24
>   *Tertiary Aliphatic Amines,* 4.25

(3d) **Primary Aromatic Side-Chain Amines.** These show most of the features of primary aliphatic amines: two or three N—H *str* bands, evolution of nitrogen with nitrous acid, and consequently no coupling with 2-naphthol. They contain an aromatic residue (detected chemically and from the infrared spectrum), but do not show aryl C—N *str*, nor do they react rapidly with bromine-water, etc.

(3e) **Primary Aryl Amines and Diamines.** These contain the group NH_2 *directly* attached to the nucleus.

All are toxic, and should be handled with care. Avoid all contact with the skin; in the event of accidental contact, wash thoroughly with soap and water.

They are easily distinguished by infrared evidence: N—H *str* (two or three bands), N—H *def*, and C—H *str* are all strong bands.

Mono-amines all give azo dyes by the diazo coupling reaction with 2-naphthol, and decolourize bromine-water and MnO_4^-. o-Diamines, when treated with nitrous acid in dilute hydrochloric acid solution, give cyclic diazoamino derivatives (yellow); m-diamines couple inter-molecularly to give complex azo dyes (e.g., Bismarck Brown); p-diamines usually behave like mono-amines. All diamines are reducing agents, and reduce Tollens' reagent rapidly; they darken considerably in air.

> Derivatives: page 71
> Tables: Primary Aromatic Side-Chain Amines, 4.26
>   Primary Aryl Amines and Diamines, 4.27

(3f) **Secondary Aromatic Amines.** This includes any compound containing a secondary NH group and an aromatic residue, whether the NH group is on the nucleus or side-chain. From the infrared spectrum, any compound showing a single N—H str peak and evidence of aromatic character is classified here. If the spectrum of a primary amine is recorded at fast scan as a liquid film, poor resolution may not show the two N—H str bands clearly; any inflection or excessive broadening of the N—H str band may indicate the need to rerun the spectrum. Commercial samples of tertiary amines (e.g., quinoline) may be wet, and show water absorption which might be construed as broad N—H str.

Most aromatic secondary amines give oily N-nitroso derivatives with nitrous acid.

(3g) **Tertiary Aromatic Amines.** All tertiary amines with an aromatic system are included here; heterocyclic bases such as pyridine and quino-line are also in this group.

No N—H str absorptions should be shown in the infrared, but normal samples may show absorption around 3500 cm^{-1} (see note in 3f).

Simple derivatives of pyridine and quinoline should be readily recog-nized by smell. Simple aryl tertiary amines (e.g., N,N-dimethylaniline) give intense green p-nitroso derivatives when treated with nitrous acid and the solution subsequently made alkaline. The para-position must be free.

> Derivatives: page 71
> Tables: Secondary Aromatic Amines, 4.28
>   Tertiary Aromatic Amines, 4.29

(3h) **Hydrazines and Semicarbazides.** Only a few members of this class are commonly encountered. Phenylhydrazine and many of its derivatives are extremely toxic: handle with care, and wash any spillage on the skin with soap and water.

As indicated above, they have similar spectroscopic properties to the amines: if primary NH_2 is present, two or three N—H str bands: if primary NH_2 is substituted, one N—H str band. Because of the number of H atoms attached to N, the N—H str bands may be quite strong. The

C=O *str* band of semicarbazides may be difficult to identify in the presence of the strong N—H *def* band around 1650 cm^{-1}.

Chemically, they are characterized by their marked reducing properties (MnO_4^-, Tollens' reagent), and by the ease with which they condense with aldehydes and ketones (see derivatives). Nitrophenylhydrazines are red.

> *Derivatives:*   page 74
> *Table:*   *Hydrazines and Semicarbazides*, 4.30

(3i) **Imines (including Schiff's Bases) and Aldehyde-ammonias.** The common feature linking these classes together is the ease with which they are converted to the carbonyl compound from which they are (hypothetically) derived; thus they all give a positive 2,4-D.N.P. test under the conditions specified in 1.12. Additionally, aldehyde-ammonias smell strongly of ammonia: simple imines are rapidly hydrolysed to give ammonia in dilute alkali (1.7): substituted imines (Schiff's Bases) give amines under these circumstances.

In the infrared, the only valuable feature is the C=N *str* absorption shown on the charts, but this may not be easily identified in aromatic compounds which absorb in this region.

> *Derivatives:*   page 74
> *Table:*   *Imines (including Schiff's Bases) and Aldehyde-ammonias*, 4.31

Nitro Derivatives of Classes 3a-3i. Preliminary indication of a nitro group may be the colour of the compound: nitro amines are commonly yellow to orange. Infrared evidence is usually sufficient to establish the presence of aryl NO_2. Note that nitro amines are much less basic than simple amines, and a few complications arise in the preparation of derivatives; they are not listed in a separate set of tables, but are found in the appropriate table for each of the classes 3a–3i.

Neutral and Amphoteric Compounds
(3j) Amides (Primary)
(3k) N-Substituted Amides (Secondary and Tertiary)
(3l) Ammonium Salts of Carboxylic Acids
(3m) Aminophenols
(3n) Amino Acids
(3o) Nitriles and Isonitriles
(3p) Azo Compounds
(3q) Nitroso Compounds
       Nitro Derivatives of Neutral Class 2 Compounds (e.g., 2e)
       Nitro Derivatives of Classes 3j–3o

Neutral compounds do not show any marked increase in solubility in acid solution or in alkali compared to water. They are distinguished in principle from amphoteric compounds, which show acidic *and* basic properties (i.e., they may be insoluble in water, but soluble in dilute acid *and* alkali). Examine all these compounds together for chemical and spectroscopic evidence of phenolic groups (page 18) or carboxylic acid groups (page 16). Such compounds are likely to be aminophenols or amino acids respectively. Note that amino acids with aliphatic or side-chain NH_2 exist as zwitterions, and show infrared absorptions due to NH_3^+ and CO_2^-.

Further classification is based on the following. All amides show $C{=}O$ *str*; nitriles and isonitriles show $C{\equiv}N$ *str* around 2150 cm^{-1}; ammonium salts show strong absorptions due to NH_4^+ ($N{-}H$ *str* around 3100 cm^{-1}) and CO_2^- (around 1350 and 1600 cm^{-1}).

Any compound *not* showing one or other of these absorptions should be examined as a possible triaryl amine (these are very weakly basic and might mistakenly be classified as neutral) or as a nitro derivative of a class 2 compound.

(3j) **Primary Amides.** These are confirmed from very definite chemical and spectroscopic evidence:

They give off ammonia when treated with hot alkali (1.7). They show $N{-}H$ *str* bands: as for primary amines, two bands (or more) are commonly seen, the separation between them being fairly large (*ca.* 170 cm^{-1}). In dilute solution, the frequencies are much higher than in solid samples; the values shown on the charts are for the hydrogen-bonded $N{-}H$ *str* absorptions seen in mulls or KBr discs.

Primary amides show a *pair* of bands around 1650 cm^{-1}. These are coupled vibrations (mainly associated with $C{=}O$ *str* and $N{-}H$ *def*) and are normally simply assigned as the Amide I and II bands. These two bands should show a separation of ca. 25 cm^{-1}, but in solid samples this value may be considerably reduced, giving the appearance of a single broad band.

Urea derivatives and other amides of carbonic acids (e.g., urethanes, barbiturates) can be treated as amides or imides, and may be identified by their infrared spectra and from physical data (m.p., etc.). Many of these are acidic, and will be classified accordingly. In many cases enol forms are favoured, and this should be borne in mind when considering the chemical and spectroscopic evidence.

Chemically, the best test is to hydrolyse with hot alkali, cool the solution, then acidify slowly with dilute acid: CO_2 is liberated. Also, provided one $CONH_2$ group is unsubstituted, ammonia will be given off with alkali, and the infrared spectrum will show the distinguishing features of $CONH_2$ discussed above.

Derivatives: page 74
Tables: *Primary Aliphatic Amides*, 4.32
               Primary Aromatic Amides, 4.33

(3k) N-Substituted Amides (Secondary and Tertiary Amides). These are fairly unreactive chemically, and this is perhaps their most useful distinguishing feature—any neutral, unreactive, nitrogen-containing compound should be considered a possible member of this class. Unlike primary amides, they do not give off ammonia with hot alkali, and are only satisfactorily hydrolysed with strong acid (e.g., 70 per cent sulphuric acid).

Spectroscopically, they are confirmed with reasonable certainty from the charts. Note that anilides have rather higher $C=O$ *str* frequencies (1680–1700 cm^{-1}) than amides derived from aliphatic amines. The origin of the band around 1540 cm^{-1} in secondary amides is not definitely known; it is associated with N—H *def* and C—N *str*. Note also that tertiary amides ($RCONR_2$) have no N—H *str* absorption.

Derivatives: page 75
Table: *N-Substituted Amides*, 4.34

(3l) Ammonium Salts of Carboxylic Acids. Ammonium salts of carboxylic acids are always easy to confirm both chemically and spectroscopically. In the infrared spectrum, the N—H *str* of $NH_4{}^+$ and the asymmetric $C\cdots O$ *str* of $CO_2{}^-$ are easiest to assign: the symmetric $C\cdots O$ *str* is more variable, but should also be sought.

Hydrolysis with warm or even cold alkali gives off ammonia; treatment of the salt (or of the alkaline hydrolysate) with dilute HCl liberates the free acid. Aromatic acids are in this way usually precipitated immediately or on cooling; the lower aliphatic acids can often be detected by smell, and extracted with ether.

Derivatives: page 76
Tables: *Carboxylic Acids*, 4.17 and 4.18

(3m) Aminophenols. These behave more or less with an amalgam of amino and phenolic properties; the appropriate tests for these groups should be carried out (3e, 2g).

In particular, aminophenols are particularly sensitive to bromine-water (easy substitution) and to $MnO_4{}^-$. They are easily oxidized by Ag^+, and hence give a positive silver mirror test (Tollens' test). They are usually sparingly soluble in water, but freely soluble in dilute acid or alkali; most are slightly coloured because of air oxidation, and a few may be quite dark unless freshly purified.

Derivatives: page 76
Table: *Aminophenols*, 4.35

(3n) **Amino Acids.** Aromatic amino acids which have the amino group directly attached to the nucleus contain the *free* carboxyl and amino groups and show most of the expected properties of these (see 3e, 2f, etc.). *o*-Amino acids are likely to be chelated and show C=O *str* around 1660 cm^{-1}.

Amino acids containing an aliphatic or side-chain amino group exist as zwitterions, and hence show $NH_3{}^+$ and $CO_2{}^-$ absorptions. All are high melting because of their salt-like structures. Most common amino acids of this class are neutral, but dicarboxylic acids will be acidic, while diamino compounds will be basic. Where a second NH_2 or CO_2H group is present in the molecule, this will show in the infrared spectrum as *free* NH_2 or CO_2H. (See 8.11.)

A fairly specific test for all α-amino acids is the ninhydrin test: to a dilute aqueous solution of the unknown, add a few drops of ninhydrin reagent (0·2 per cent)—a deep blue or blue-green colour develops. Check the infrared spectrum for the other absorptions shown by typical α-amino acids.

> *Derivatives:*   page 76
> *Table:*   *Amino Acids*, 4.36

(3o) **Nitriles and Isonitriles.** The infrared spectrum of these compounds is usually sufficient to identify them: very few absorptions other than C≡N *str* occur around 2150 cm^{-1}.

Nitriles give off ammonia very slowly with hot alkali; common isonitriles have intense, repulsive odours, but are so highly toxic that they are unlikely to be supplied to students as unknown compounds.

> *Derivatives:*   page 77
> *Table:*   *Nitriles*, 4.37

(3p) **Azo Compounds.** Most common azo compounds are produced by the diazo coupling reaction, and contain a phenol or amino group as well as the azo group. All are highly coloured, and yield colourless products if reduced by tin and hydrochloric acid. Good evidence for the azo group is obtained from the electronic absorption spectrum (see page 147). Infrared evidence is not very helpful.

> *Derivatives:*   page 77
> *Table:*   *Azo Compounds*, 4.38

(3q) **Nitroso Compounds.** The commonest nitroso compounds are *p*-nitroso-N,N-dimethylaniline derivatives, and these are intense dark green compounds. Since these are also tertiary amines, they form salts with acid (though slowly) and these salts are deep red. Infrared evidence is not normally of much diagnostic value.

Nitrosophenols exist in the form of the isomeric quinone monoximes, and consequently show O—H *str* absorptions around 3500 cm^{-1}, and C=O *str* bands around 1640 cm^{-1}.

Derivatives: page 78
Table: *Nitroso Compounds*, 4.39

Nitro Compounds. Neutral nitro compounds may be nitro derivatives of class 3 or class 2 compounds. Class 3 compounds should already have been detected. Class 2 functions can be detected as outlined there, with the exception of *nitro hydrocarbons* and *nitro ethers*. Treat these as follows.

For aromatic compounds, search the infrared spectrum for NO$_2$ absorptions. Reflux a small amount of the unknown in a test-tube with a piece of granulated tin and dilute hydrochloric acid: the resulting solution should contain the salt of a primary amine. Test for primary amine by diazotization and coupling (see page 21). (See also 2e.)

Derivatives: page 78
Table: *Aromatic Nitro Hydrocarbons and Nitro Ethers*, 4.40

Simple nitro alkanes are best identified by infrared evidence for NO$_2$, and by boiling point.

Table: *Nitro Alkanes*, 4.41

Acidic Compounds
(3r) Imides
(3n) Some Amino Acids
     Nitro Derivatives of these and of acidic Class 2 Compounds

(3r) Imides, Acidic Urea Derivatives, Barbiturates. Imides differ from amides mainly in the acidity of the NH group: spectroscopically and chemically they are very similar indeed, and they are best examined and treated as discussed there (3j, 3k). They show only one N—H *str* band (CONHCO); the pair of bands around 1650 cm^{-1} tend to be more separated than in the case of primary amides, although this is much more clearly seen in solution spectra.

In the special case of cyclic urea derivatives (e.g., barbiturates), these are acidic by virtue of their enolic nature, and their infrared spectra show evidence of this (bonded O—H *str*). Barbituric acids have distinctive electronic absorption spectra; see table 6.4.

Derivatives: page 78
Table: *Imides and Cyclic Urea Derivatives*, 4.42

(3n) Acido Amino Acids. These are discussed fully above, under 3n.

Nitro Compounds. Acidic nitro compounds are likely to be nitro carboxylic acids or nitrophenols, but first check for the presence of imide as in 3r.
Nitro Carboxylic Acids behave as normal carboxylic acids, so that they

will be distinguished by the characteristic infrared spectrum of CO_2H (2f). The nitro group can be confirmed from the spectrum also. Thereafter, treat these compounds as if they were normal carboxylic acids. They are listed at the end of table 4.18.

Nitrophenols, however, may be significantly different from simple phenols: they are much more acidic, and may liberate CO_2 from bicarbonate solution. They dissolve to some extent in water giving yellow solutions, but addition of alkali gives a much more intensely yellow solution due to the formation of the nitrophenate ion. Acidification of this alkaline solution reduces the colour intensity again.

  Derivatives: page 70 (with phenols)
  Table: *Nitrophenols*, 4.20 (with phenols)

2.4 Compounds containing C, H, Halogen, and possibly O

(4a) Acyl Halides
(4b) Alkyl Halides and Aromatic Side-Chain Halides
(4c) Aryl Halides (nuclear halogen)
   Halogen Derivatives of Class 2 Compounds

 If the compound is pungent-smelling, and reacts exothermically (or even violently) with water to give halide ion, it is almost certainly an acyl halide: confirm this below (4a).

 Otherwise, examine the infrared spectrum of the compound for C=O *str* and O—H *str*; if either of these is present, the compound is a halogen derivative of one of the following class 2 compounds, and should be examined as such: carboxylic acid or anhydride, phenol or enol, aldehyde or ketone, ester, or alcohol. (Impure samples of acyl halides may contain free CO_2H, and this will show in the infrared spectrum.) Tables of chloro acids follow simple acids, chlorophenols follow simple phenols, etc.

(4a) **Acyl Halides.** These are always acidic because of easy hydrolysis to the acid and hydrogen halide. Lower aliphatics are pungent-smelling, lachrymatory liquids, which react vigorously with water; aromatic members have similar physiological effects, but react more slowly with water. (Side-chain aromatic members behave as aliphatics.)

 They all show strong C=O *str* absorptions at relatively high frequencies (see charts), which makes them easy to identify. Aryl members appear at the low-frequency end of the range; benzoyl chloride itself shows two bands (1773 and 1736 cm^{-1}), the second being of unknown origin.

  Derivatives: page 79
  Table: *Acyl Halides*, 4.43

(4b) **Alkyl Halides and Aromatic Side-Chain Halides.** Most of these are liquids, even those with relatively high molecular weight. The halogen atom is not usually hydrolysable with water alone (aliphatic iodine is exceptional) so that they react only with ethanolic $AgNO_3$, but not with aqueous. Otherwise they are fairly inert. Benzylic halides are exceptional (see below).

Gem-dihalogen compounds are frequently more easily hydrolysed, giving an aldehyde (terminal —CHX_2) or ketone (—CX_2—). Tertiary halides (together with allylic and benzylic halides) are also exceptional in the ease with which they hydrolyse. Primary halides are least reactive.

Spectroscopically, they show few changes from the parent hydrocarbon, unless the spectrum can be scanned below 600 cm^{-1}, where many carbon–halogen vibrations are found. Polyhalogen compounds absorb at higher frequencies. An interesting case is fluorinated material, where the extremely intense C—F *str* (1000–1400 cm^{-1}) dominates the spectrum: fluorine compounds are not otherwise considered in this book.

> *Derivatives:*   page 80
> *Table:*   *Alkyl Halides and Aromatic Side-Chain Halides*, 4.44

(4c) **Aryl Halides.** With halogen directly attached to the nucleus, and consequently unreactive, the simplest method of dealing with these is to treat them as hydrocarbons (2e); spectroscopically, they are indistinguishable from these above 600 cm^{-1}, and only aryl chlorides show absorption above 400 cm^{-1} associated with C—Cl *str*.

> *Derivatives:*   page 81
> *Table:*   *Aryl Halides*, 4.45

2.5   Compounds containing C, H, S, and possibly O

(5a) Mercaptans (Thiols), Sulphides (Thioethers), Disulphides, and Thioacids
(5b) Sulphonic Acids
(5c) Sulphonate Esters
(5d) Sulphate Esters

Initial distinctions among these classes is based on the following: all class 5a compounds have extremely strong, unpleasant smells; sulphonic acids are relatively odourless, but strongly acidic, and commonly are hygroscopic liquids or solids; sulphonate esters are comparatively inert; sulphate esters hydrolyse rapidly in water to give ionic sulphate.

(5a) **Mercaptans, Sulphides, Disulphides, and Thioacids.** All of these compounds have strong unpleasant smells, and it remains only for infrared evidence to identify the S—H *str* absorption in mercaptans (including thiophenols) and thioacids. Thioacids contain the C=S group and have other infrared features as shown on the charts. Thiophen derivatives are included here.

Derivatives: page 81
Table: *Mercaptans, Sulphides, Disulphides, and Thioacids,* 4.46

(5b) **Sulphonic Acids.** Common aliphatic acids are liquids, while the aromatics are solids. All are hygroscopic, readily soluble in water, and react vigorously with bicarbonate. They are frequently met in the form of their salts (class 11). Infrared information is fairly conclusive in identification.

Derivatives: page 82
Table: *Sulphonic Acids,* 4.47

(5c) **Sulphonate Esters.** These are fairly unreactive, and you may have to lean heavily on infrared identification. The most unambiguous proof of identity is the hydrolysis to a sulphonic acid with aqueous alkali, followed by the identification of this acid by (e.g.) its S-benzylthiuronium salt.

Derivatives: page 83
Table: *Sulphonate Esters,* 4.48

(5d) **Sulphate Esters.** The dialkyl or diaryl sulphates are more difficult to identify than the sulphonate esters, and may be confused with these. They can be hydrolysed to sulphuric acid and an alcohol (or phenol); the sulphuric acid can be confirmed by the $BaCl_2$ test, and the alcohol or phenol isolated and identified.

Monalkyl and monoaryl sulphates have a free acidic hydrogen; they are most frequently encountered as their metal salts, for example potassium ethyl sulphate, sodium dodecyl sulphate. As for dialkyl and diaryl sulphates, they should be hydrolysed to sulphuric acid and alcohol (or phenol).

Derivatives: page 83
Table: *Sulphate Esters,* 4.49

2.6   Compounds containing C, H, N, Halogen, and possibly O

(6a) Halogen-Substituted Nitro Hydrocarbons and Ethers
(6b) Acyl Halides of Nitro Carboxylic Acids
(6c) Amine Hydrohalides
(6d) N-Halogeno Compounds
       Halogen Derivatives of Class 3 Compounds

All of the compounds in 6a–6d (excepting a few in 6a) have easily hydrolysable halogen atoms. 6a Compounds are neutral, coloured, (yellow or orange), and show NO_2 absorptions in the infrared. Compounds in classes 6b, 6c, and 6d all react with water to give acidic solutions. None shows O—H *str* in the infrared.

Any compound *not* thus described (i.e., compounds with unreactive halogen, neutral colourless compounds, etc.) must be treated as simple halogen derivatives of class 3 compounds (e.g., chloro amines, chloro-nitro carboxylic acids, etc.).

(6a) Halogen-Substituted Nitro Hydrocarbons and Ethers. Any compound which is entirely aliphatic is outside the present discussion; only aromatic compounds with halogen and nitro directly attached to the nucleus will be considered (nitro derivatives of side-chain halides are usually powerfully lachrymatory). With the exception of *m*-nitrohalogeno compounds, the halogen atom is replaceable by nucleophilic substitution: mononitro compounds may need hot alkali, trinitro compounds hydrolyse in water. It is usually easy to convert them (for derivative purposes) to the corresponding phenol by hydrolysis. Infrared analysis confirms the presence of NO_2

> *Derivatives:*   page 83
> *Table:*   *Halogen-Substituted Nitro Hydrocarbons and Ethers,* 4.40a

(6b) Acyl Halides of Nitro Carboxylic Acids. Again, purely aliphatic compounds are outside the present discussion; the nitro group must be attached to an aromatic nucleus, although the acyl halide group may be on the nucleus or on the side-chain.

They can be identified as acyl halides using the notes given in (4a). They can then be hydrolysed to the corresponding aromatic nitro acid by refluxing with water: if the acid does not crystallize out on cooling, extract with a little ether. Identify the acid as if it were a member of class 2f. Repeat the sodium fusion test on the free acid; a second, inert halogen atom may also be present on the nucleus.

> *Derivatives:*   page 79 (as for class 4a)
> *Table:*   *Acyl Halides of Nitro Carboxylic Acids,* 4.43

(6c) Amine Hydrohalides. Within this category are the hydrohalide salts of all basic nitrogen compounds, including simple amines (primary, secondary, or tertiary), heterocyclic bases, hydrazines, semicarbazides, and amino acids.

All are quite strongly acidic, have ionic halogen (aqueous $AgNO_3$ reacts), and on treatment with dilute alkali liberate the free base—often as an oil unless it is water soluble. In the infrared, absorptions due to NH^+, NH_2^+, and NH_3^+ are present (see charts, 6c); these are quite different from the infrared absorptions of the free base. Hydrazine hydrohalides, etc., show absorptions due to NH_3^+ *and* non-ionic NH.

Isolate the free base and treat it as a class 3 compound; repeat the sodium fusion test on the free base: a second, inert halogen atom may be present.

In the particular case of amino acid hydrohalides, these show the infrared characteristics of $NH_3{}^+$, etc., but also of *free* CO_2H, so that identification is relatively easy. For derivatives of these, treat as amino acids (3n).

For all class 6c compounds, identification of the free base is carried out as indicated in the appropriate class 3 section: derivatives and tables are as given there.

(6d) N-Halogeno Compounds. The commonest examples are N-bromo imides and N-chloro imides. These liberate molecular halogen in aqueous solution, and this can be detected by starch-iodide paper. Hydrolyse to the halogen-free derivative, and identify this as a completely new compound, repeating the sodium fusion test, etc. Note that hydrolysis of N-halogeno imides will give the corresponding dicarboxylic acid (e.g., succinic acid) with loss of nitrogen as ammonia. Hydrolyse as for amides, page 74.

Table: *N-Halogeno Compounds*, 4.50

2.7   Compounds containing C, H, N, S, and possibly O

(7a) Ammonium Sulphonates
(7b) Thioureas and Thioamides
(7c) Sulphonamides and N-Substituted Sulphonamides
(7d) Aminosulphonic Acids and Derivatives
(7e) Amine Sulphates
       Nitro Derivatives of these and Class 5 Compounds

No simple set of tests or correlation charts will distinguish among these groups, but the following evidence will suffice in most cases. Nitro groups do not interfere with any of the classification tests given below: as usual the presence of NO_2 may be initially inferred from colour, and confirmed spectroscopically and from the m.p. tables, etc.

Ammonium salts, together with thioamides and thioureas which still contain an unsubstituted $CSNH_2$ group, give off ammonia when treated with cold or hot alkali. Simple (primary) sulphonamides give off ammonia only with soda-lime: N-substituted sulphonamides give amines.

Aminosulphonic acids are dipolar salts (zwitterions), and therefore do not melt; they show acidic but *not* basic properties. Primary amino groups attached to the aromatic nucleus in these compounds can be diazotized, and subsequently coupled with alkaline 2-naphthol. Amine sulphates are easily confused with aminosulphonic acids, but the former give a positive test for ionic sulphate ($BaCl_2$).

(7a) Ammonium Sulphonates. These are usually neutral or slightly acidic, soluble in water to a considerable extent, and give off ammonia with cold or hot alkali. They do not give a precipitate of $BaSO_4$ when treated with $BaCl_2$. No clear confirmation may be obtained from the infrared spectrum, because of the many points of similarity with the spectra of other sulpho compounds (but note the absence of strong absorption around 1400 cm^{-1} which distinguishes ionic from covalent sulphonate). Look for NH_4^+ *str* (3100–3200 cm^{-1}) and the sulphonate absorptions shown on the charts (under sulphonic acids, 5b).

The best identification methods are (i) attempt to prepare an S-benzyl-thiuronium salt directly: this will confirm ionic sulphonate and serve as a derivative, and (ii) boil a small amount (0·2 g) with dilute alkali (3 cm^3) until no more ammonia is given off, acidify with dilute hydrochloric acid, and evaporate the residue to dryness. Examine the residue as a possible sulphonic acid (5b); make allowances for possible impurities (sodium chloride, and a little residual HCl).

   Derivatives: page 82 (as for sulphonic acids)
   Table: *Sulphonic Acids*, 4.47

(7b) **Thioureas and Thioamides.** If free $CSNH_2$ is present, these hydrolyse like normal ureas and amides to give off ammonia. Otherwise they hydrolyse to an amine which must be isolated and identified. Infrared evidence for the C=S group is reasonably easy to ascertain. This appears as a strong band, and may lie close to the C=N absorption around 1300 cm^{-1} (*m*); the intensity difference is usually sufficient to differentiate, but note that thiourea itself has its C=S *str* absorption at very high frequency (1410 cm^{-1}). Primary $CSNH_2$ groups show multiple N—H *str* absorptions.

   Derivatives: page 84
   Table: *Thioureas and Thioamides*, 4.51

(7c) **Sulphonamides and N-Substituted Sulphonamides.** The groups SO_2NH_2 and SO_2NH are detected from infrared evidence, and from the alkali solubility of the compounds containing them. A few are very weakly acidic, and may show little solubility in alkali; confusion with N,N-disubstituted sulphonamide ($SO_2N<$) should be resolved by the presence of N—H *str* bands.

N,N-Disubstituted sulphonamides are less easily confirmed; they are neutral, crystalline materials with sharp melting points, and with soda-lime they decompose to the constituent sulphonic acid and secondary amine, which must be isolated and identified separately.

In all cases, hydrolysis to and identification of the sulphonic acid is the best confirmation. This is discussed under the preparation of derivatives.

   Derivatives: page 84
   Table: *Sulphonamides*, 4.47

(7d) Aminosulphonic Acids and Derivatives. Aminosulphonic acids with an aryl primary amine group (the commonest examples) have the following distinguishing features. The infrared spectrum shows NH_3^+ and ionic sulphonate absorptions, since they exist exclusively as zwitterions. They decompose before melting (ionic crystal forces) and dissolve to some extent in water, especially hot water, to give acidic solutions. They liberate CO_2 from bicarbonate, and can be diazotized as follows: they are dissolved in aqueous Na_2CO_3, then $NaNO_2$ is added, the temperature of the solution being maintained around 10°; the solution is made just acid by the slow addition of dilute hydrochloric acid. On pouring into alkaline 2-naphthol, a *soluble* ($SO_3^-Na^+$) red azo dye is formed.

Functional derivatives of the sulpho group (sulphonamide, sulphonate ester, etc.) have the amino group present as *free* NH_2, and therefore show typical chemical and spectroscopic behaviour for primary aryl amines (3e) and for these other groups (7c, 5c, etc.).

Aminosulphonic acids are often encountered as their salts with alkali metals (class 11). These show ionic sulphonate, but free amine absorptions in the infrared.

Derivatives: page 85
Table: *Aminosulphonic Acids*, 4.47a

(7e) Amine Sulphates. This group includes the sulphates (more often the hydrogen sulphates) of all organic bases, including heterocyclics. They are acidic salts, and with dilute alkali the free amine is liberated (often as oily droplets for liquid amines or low-melting solid amines). When treated with dilute hydrochloric acid and $BaCl_2$ solution, they give $BaSO_4$; this distinguishes them from aminosulphonic acids, many of whose reactions and spectroscopic features they share (acidity, solubility, diazotization if the amine is primary aryl). In the infrared spectra, although the correlation charts show distinctions between amine sulphates and aminosulphonic acids, confident distinction is not always possible by this means. Isolation and characterization of the free amine is the most certain method of identification.

Derivatives: page 71 (as for amines)
Tables: *Amines*, 4.23–4.30

2.8 Compounds containing C, H, S, Halogen, and possibly O

(8a) Sulphonyl Halides
Halogen-Substituted Class 5 Compounds

(8a) Sulphonyl Halides. Test the reactivity of the halogen atom: in the SO_2X group it can be hydrolysed off by boiling for a few minutes with water—add nitric acid and $AgNO_3$ to test for halide ion. Check also the infrared features, which are usually conclusive. If any further confirmation is necessary, attempt the preparation of an anilide derivative as for toluene-*p*-sulphonyl chloride (see derivatives of amines, page 72).

Derivatives: page 82 (as for sulphonic acids)
Table: *Sulphonyl Halides*, 4.47

Halogen-Substituted Class 5 Compounds. If the halogen atom is unreactive, treat the compound as a simple class 5 compound. If the halogen is reactive, e.g., to aqueous reagents, then more than one functional group must be identified from class 4, class 5, or class 8a. No simple subdivision of these possibilities can be achieved: it is necessary to work systematically through these functional groups until confirmation of identity is obtained. Derivatives and tables are incorporated into the appropriate section for each functional group.

2.9   Compounds containing C, H, N, Halogen, S, and possibly O

(9a) N-Halogenosulphonamides
(9b) Hydrohalide Salts of Thiourea and Derivatives
      Any combination of the Classes 2–9

We need distinguish only whether the compound belongs to either class 9a or 9b; if not, then a systematic search must be made of all other functional classes, starting with class 8, and working backwards.

(9a) **N-Halogenosulphonamides.** Usually these are encountered as the N-chloro or N-bromo derivatives, and often as their sodium salts. The halogen is released in water as molecular halogen, and can be detected with starch–iodide paper.

They are converted to the corresponding sulphonamides by boiling (0·5 g) with dilute alkali and '20 volume' hydrogen peroxide (10 cm^3 of each): acidify with concentrated hydrochloric acid, boil until the liquor is clear, then filter off and recrystallize the product from aqueous ethanol. Examine this as a sulphonamide (7c).

Derivatives: as for Sulphonamides, page 84
Table: *N-Halogenosulphonamides*, 4.52

(9b) **Hydrohalides of Thioureas.** The commonest example of this class is S-benzylthiuronium chloride. Unlike class 9a compounds, these release only ionic halide ion in aqueous solution. They decompose on heating, particularly in alkali, to give the corresponding mercaptan (S-benzyl-thiuronium chloride gives benzyl mercaptan, $PhCH_2SH$); these are recognizable by their disagreeable smells. They form salts with carboxylic and sulphonic acids (q.v.), and form charge transfer complexes with picric acid.

Table: Hydrohalides of Thioureas, 4.53

2.10   Compounds containing C, H, P, and possibly O

(10a) Phosphate Esters
(10b) Phosphite Esters

Although many other organophosphorus compounds are known, these esters are the most important commercially (as plasticizers, fuel additives, and insecticides).

They can both be identified spectroscopically, the absence of $P{=}O$ absorption being characteristic of trisubstituted phosphites. Acid esters are not included in this category.

For identification of individual members, they are hydrolysed in boiling alkali to phosphoric or phosphorous acid and the corresponding OH compound (alcohol or phenol). The OH compound must then be isolated and identified. (Test the hydrolysate to confirm the characterization of phosphate or phosphite ions). Hydrolyse as for carboxylic esters, page 60.

Derivatives: As for Alcohols (2d) and Phenols (2h)
Table: Phosphate and Phosphite Esters, 4.54

2.11   Compounds containing a Metal

The presence of a metal ion will be indicated by an incombustible residue when the compound is burnt. This residue should be treated with ammonium nitrate or concentrated nitric acid and again strongly heated; repeat this process until all carbon has been burnt off, then identify the metal ion by standard inorganic procedures.

It is difficult to be specific about the best method of identifying a metal-containing organic compound, but the following guide lines will normally be sufficient.

Remember that the infrared spectrum of a metal derivative of an organic functional group will usually be quite different from that of the free group, so that carboxylate ions absorb differently from carboxyl groups, etc. Sulphonates, however, have similar spectra to sulphonic acids, with the exception of O—H absorptions.

(11a) **Compound contains a Metal and S.** Possibly a *metal sulphonate*, the *bisulphite compound* of an aldehyde or ketone, or an *alkyl hydrogen sulphate*.

Metal sulphonates are neutral and water soluble (alkaline–earth salts less so); they decompose before melting. Infrared absorptions are similar to those of sulphonic acids, with the exception of O—H *str* peaks (see charts, 5b). Derivatives: as for sulphonic acids (5b).

Bisulphite compounds of aldehydes and ketones have two very characteristic reactions: (i) with dilute sulphuric acid, they decompose giving off SO_2 and (ii) they give the 2,4-D.N.P. test for the carbonyl compound. Identification is based on this latter reaction; prepare the 2,4-D.N.P. derivative and thereby identify the aldehyde or ketone (2a). Should any ambiguity still remain, a second derivative (e.g., the *p*-nitrophenylhydrazone) can be prepared in a like manner.

Alkyl hydrogen sulphates: the alkali metal salts are most commonly met, particularly of long-chain alcohols, as these are used as light-duty detergents (shampoos, etc.). In the infrared, they differ somewhat from sulphate esters (see charts, 5). These alkali metal salts are water soluble, but heavy metal salts are less so. To identify, hydrolyse by refluxing with dilute hydrochloric acid; isolate and identify the alcohol (2d).

(11b) **Compound contains a Metal and N.** Most likely the salt of an *imide*; treat with acid and try to isolate the free imide (3r). The imide may undergo further hydrolysis, with loss of ammonia, to the dicarboxylic acid (2f).

(11c) **Compound contains a Metal and N and S.** These are possibly the salts of *sulphonamides, aminosulphonic acids*, or a *sulphoimide* such as saccharin. The most consistently successful method of investigation is to treat them with acid and try to isolate the corresponding sulphonamide, etc. Failing this, sulphonamides and sulphoimides can be hydrolysed to the nitrogen-free derivative, which can then be isolated and identified.

(11d) **Compound contains a Metal, but no N, S, or Halogen.** These are most likely to be salts of carboxylic acids: salts of phenols, enols, and alcohols are also possible, but these are strongly basic in reaction, alkoxides particularly being very strong bases. Acidify to liberate the free acid, phenol, alcohol, etc., and then identify this separately. The infrared spectra of these salts may be quite different from that of the conjugate acid, as discussed above for carboxylates. Salts of dicarboxylic acids may be the acid salts, and contain free carboxyl together with carboxylate ion. Hydroxy acids have salts which show the O—H *str* of the OH group. Many carboxylate salts crystallize with water of crystallization, which will show in the infrared spectrum as broad O—H *str* around 3500 cm^{-1}.

(11e) **All Others.** Any compound which cannot be identified under one or other of the above sections should be treated with acid; the organic residue should be isolated and treated as a new unknown, noting that the acidification may have produced unexpected changes in the parent structure.

3

The Infrared Spectrum

It is assumed that the student has had a course of instruction on the theory of infrared absorption by organic molecules. The following practical notes are apposite only to the identification of simple functional groups by infrared spectroscopy, which is a very small part of the subject. Suitable texts to use in conjunction with this scheme of qualitative organic analysis are listed here.

The major reference work for all infrared studies is L. J. Bellamy, *The Infra-red Spectra of Complex Molecules*, Methuen and Co., London, 2nd Ed., 1958. It is hoped that all students will have access to a copy of this book. The same author and publisher have published a second volume (*Advances in Infrared Group Frequencies*) which surveys more recent information.

Less comprehensive and, in consequence, simpler coverage is provided by a number of texts, including the following.

D. H. Williams and I. Fleming, *Spectroscopic Methods in Organic Chemistry*, McGraw-Hill Publishing Co. Ltd., London, 1966.

J. R. Dyer, *Applications of Absorption Spectroscopy of Organic Compounds*, Prentice-Hall, Englewood Cliffs, N.J., 1965.

A. D. Cross, *Introduction to Practical Infrared Spectroscopy*, Butterworths, London, 2nd Ed., 1964.

J. C. P. Schwarz (ed.), *Physical Methods in Organic Chemistry*, Oliver and Boyd, Edinburgh, 1965.

W. Kemp, *Organic Spectroscopy*, Macmillan, London, 1975.

Access to a catalogue of published infrared spectra may enable the student to compare the spectra of known compounds with that of his unknown. These catalogues are also valuable in allowing a comparison to be made even between the spectra of similar compounds, so that the differences produced in the spectra can be related to differences in the molecular structure. The most comprehensive set is the D.M.S. system: *Documentation of Molecular Spectroscopy*, Butterworths, London. Information on other spectroscopic data is also included, on indexed punched cards.

Of more limited scope, and therefore in some respects more directly useful to the student, is the excellent Mecke catalogue of around 1800 spectra: Mecke and Langenbucher, *Infrared Spectra of Selected Chemical Compounds*, Heyden and Son, London, 1965.

The principal difficulty in approaching an infrared spectrum is deciding how much or how little can justifiably be deduced from it. Taken in conjunction with the chemical evidence elicited from earlier chapters, the principal aim at this stage is to ensure that no important functional group has escaped detection; most functional groups confer chemical properties on the molecule which are easily detected, but other functional groups give less definite reactions in the test-tube (e.g., alcohols, esters, or secondary amides). This chapter will be concerned with the systematic screening of the infrared spectrum for evidence of these less reactive groups, and will indicate how chemical confirmation of their presence can be obtained.

The *absence* of an absorption band from the spectrum may be a very definite indication that a particular functional group is absent from the molecule. Conversely, the *presence* of a band may be due to more than one factor, and may not arise from any expected feature in the molecule. This is especially true in the region 650–1300 cm^{-1}, where many unassigned molecular vibrations occur in the majority of organic compounds. The general rule here is: if a functional group gives rise to a *series* of bands, rather than one single band, then its presence in the molecule can usually be detected with some certainty. Thus primary aryl amines are easily identified, since the spectrum shows N—H *str*, N—H *def*, and C—N *str*, all of which are strong bands.

3.1 Resolution and Sampling

A few notes must be included here on the practical details of obtaining a good infrared survey spectrum; full details of the facilities available to the student are best described by the instructor, but the following is a series of essential check-points which the student must continually have in mind in infrared work.

Do not begin to study the infrared spectrum unless its clarity and resolution are wholly satisfactory, otherwise much misleading information may be obtained. The peaks should be clearly defined and sharp, and the strongest peaks should come near to zero per cent transmission at their absorption maxima: these strong peaks should *not* be flattened off, indicating that too much sample is being used. Be prepared to run two or even three spectra in order to obtain optimum concentration and resolution. For detailed work, more than one good spectrum should be obtained, at differing concentrations or differing film thickness, so that the strong peaks are clearly resolved in the dilute spectrum, and the weak peaks easily studied in the strong spectrum.

Liquid samples should be recorded as thin films on rock salt flats. If the design of the instrument permits it, hand scan the spectrum, with the recorder pen off the chart, to check the strength of the absorption. If the spectrum is too strong, *gently* squeeze the rock salt flats together to produce a thinner film. Handle the flats at all times by their edges, *never* on their faces. If the spectrum is too weak, add more material to the flats, and mount them loosely to ensure an adequate film. Even relatively volatile liquids (e.g., acetone) can be dealt with in this way: if evaporation is very rapid, it is usually possible to record the spectrum in two halves, adding more sample at the half-way stage. Failing this, use a liquid cell of path length 0·01–0·1 mm.

Solid samples are best recorded in KBr (or other alkali halide) discs. If equipment is not available for making discs, then the spectra should be recorded as mulls; if mulls are used, always run two spectra—one in Nujol mull (high molecular weight paraffinic oil), the other in a complementary mulling agent such as hexachlorobutadiene or Fluolube. Where mulls are used, it is essential to have the infrared spectra of the mulling agents available for consultation, and the first step in any such infrared study must be to mark on the spectra the peaks due to the mulling agent.

In making KBr discs, it should be remembered that many solid organic compounds are surprisingly hard, and therefore difficult to disperse completely in the KBr unless a commercial grinding mill is available. Poor dispersion will produce a poorly resolved spectrum, with very high background 'absorption' (due to intense scattering of the light beam). The dispersion must be improved, by regrinding the substance with the KBr; the original disc may be re-ground and re-pressed. If one regrinding is insufficient, then repeat a second time until no further improvement in the resolution is obtained.

If a sharp spectrum is still not obtained, then the dispersion of the solid in the KBr can often be improved by grinding the two together in the presence of a small amount of a suitable volatile solvent (e.g., chloroform); if the grinding is carried out under an infrared lamp the solvent will quickly evaporate, and the disc can be pressed in the usual manner.

As a last resort, the solid can be dissolved in a volatile solvent, and the solution slowly dropped on to a rock salt flat held under an infrared lamp. Evaporation of the solvent produces a thin film of crystalline material on the flat; it is not always possible to produce strong spectra in this way without impairing resolution.

Solid samples which still do not produce good spectra in these ways can only be studied in solution.

Water absorption frequently appears on the infrared spectrum, as a broad band around 3500 cm^{-1}, associated with strongly hydrogen bonded O—H *str*. The shape of this peak is characteristically that of an approximate equilateral triangle, and if such an absorption is present, then either the sample contains water, or the KBr disc in the sample beam is wet compared to that in the reference beam. It is also possible that the compound contains water of crystallization; many ionic crystalline materials do, e.g., carboxylate salts, sulphanilic acid.

Solution spectra are most commonly run in chloroform, carbon tetrachloride, or carbon disulphide solution; reference spectra for these solvents should be available for consultation. The solution (0·1–5 per cent) is introduced into the cell using a hypodermic syringe; the cell is then placed in the sample beam of the instrument, and a complementary cell containing solvent alone is placed in the reference beam. For the most accurate work, a series of sealed cells of known path length can be employed, together with a variable path length cell in the reference beam (so that solvent absorption can be carefully balanced out). For routine qualitative analysis, a pair of demountable cells with (nominally) identical path length is adequate, easier to clean, and cheaper to maintain.

Since solvent peaks cancel out between sample and reference beams, the spectra obtained are those of the pure compound.

In the spectral regions where the solvent itself absorbs strongly, insufficient light reaches the instrument detector: no reliable information about the compound can be obtained from these areas, which can only be examined by recording a second spectrum with a complementary solvent, which is transparent at these wavelengths.

Much valuable and accurate information can only be obtained from infrared studies on solutions of organic compounds; but although solution spectra will be occasionally referred to, they are generally outside the scope of this book, and it will be assumed that routine survey spectra on solid substances are obtained on KBr discs or in mulls. It is also assumed that the spectra of liquids are recorded on thin films.

Spectra obtained in the solid state or on the bulk liquid will usually differ from those of the same compound obtained in dilute solution, when intermolecular forces may be radically altered. The correlation charts show average values for band positions in the solid state or bulk liquid; where gross changes occur on passing to dilute solution, this is indicated on the charts. This is particularly so for amides, amines, alcohols, carbonyl compounds, etc. More accurate limits for solution spectra are given in Bellamy's book, which should be consulted in cases of doubt.

3.2 Examining the Spectrum

The spectrum will already have been examined continuously during the course of the chemical investigations on the compound; the following notes are intended to ensure that no important functional group is undetected.

(a) The Region 2800–4000 cm^{-1}: C—H, O—H, N—H *str*

An examination of the C—H *str* can help to distinguish aromatic from aliphatic compounds, but this has already been dealt with (page 10).

Absorption due to O—H *str* varies greatly in position and intensity depending on the degree of hydrogen bonding present. Firstly ensure that no water absorption is present (page 41).

The more strongly an OH group is hydrogen bonded, the more likely is it that the *bonding* group will be easily detected by chemical and infrared means (e.g., OH in carboxyl is strongly hydrogen bonded, but the carbonyl group is extremely easy to detect). Thus carboxyl OH and the enolic OH of β-diketones show very broad, very strong O—H *str* absorptions, often reaching from 2700–3300 cm^{-1}, with the bases of the peaks spreading out even wider.

Phenols, equally, are easily detected chemically, and the characteristic O—H *str* absorption serves to confirm.

The most likely hydroxylic function to escape chemical detection is alcoholic OH. An OH group unaffected by hydrogen bonding will only be observed in very dilute solutions, so that alcohols examined as liquid films or in KBr discs will not show the sharp absorption around 3600 cm^{-1} due to non-bonded OH. Instead, they show one medium-strong peak, usually centred around 3300 cm^{-1}. If such an absorption is present, and the compound does not contain nitrogen, then the compound should be examined carefully as a possible alcohol (class 2d).

If the compound contains nitrogen, it should be examined as follows: absorptions due to N—H *str* can be divided into two types, depending on whether one N—H *str* band or more than one band is seen.

If a strong absorption made up of two bands of almost equal intensity is present around 3300 cm^{-1}, the compound is likely to contain an NH$_2$ group. (A third band of lower intensity may be present on the low-frequency side of the doublet.) The two N—H *str* bands of primary amines differ in appearance from those of primary amides, the latter being further apart and associated with the other amide absorptions around 1670 cm^{-1} (see below). In each case, search also for the strong N—H *def* absorption around 1600–1640 cm^{-1}, noting that this is a much stronger absorption than C=C *str* and C$\equiv$C *str* (of an aromatic ring) which occur near there. Amines are usually easily detected chemically (see 3a, etc.) but evidence for basicity may occasionally be inconclusive. Primary amides are likewise easily confirmed chemically (see 3j) provided care is taken to detect the ammonia given off with hot alkali (1.7).

In nitrogen-containing compounds, if only one weak-medium band is shown around 3300 cm^{-1}, the compound will probably be a secondary amine or secondary amide. Secondary amines vary widely in their basicity, and hence also in the ease with which they can be detected; the most likely to be missed chemically are the diaryl amines (3f). If there is any doubt about the basicity of the compound, repeat carefully the solubility tests in chapter 1 (1.5, 1.6, 1.7, and 1.9). Secondary amides (3k) are difficult to detect chemically; look for other characteristic amide absorptions indicated below and on the charts.

(b) The Regions 2200 cm^{-1} and 1300–1600 cm^{-1}

If the compound contains nitrogen, a quick inspection for CN and NO_2 is worthwhile.

The strong C≡N *str* absorption is distinctly seen around 2200 cm^{-1} (see charts, 3o), but nitriles as a class can easily escape chemical detection. Nitro groups are also difficult to detect chemically by unambiguous methods, but the assignment of the NO_2 absorptions is usually straightforward (see charts, just above 3a); occasionally the presence of a number of other strong bands around the NO_2 absorptions makes it more difficult.

(c) The Region 1600–1750 cm^{-1}: C=O *str*

This absorption is one of the most carefully studied bands in the infrared spectrum. For the present consideration, we can ignore the detection of carboxylic acids and their anhydrides, as these will certainly have been detected by the processes of chapter 2. If a strong absorption is shown in this region, and the compound does not contain nitrogen, then it may be an aldehyde, a ketone, or an ester. If one or other of these functions has not been detected in chapter 2, repeat the 2,4-D.N.P. test (1.12). If this test is still negative, then the compound is likely to be an ester, and should be confirmed as such by the chemical and spectroscopic consideration outlined in chapter 2 and on the charts.

If the compound contains nitrogen and strong absorption is present around 1650 cm^{-1}, the compound may be an amide. Primary amides should have been detected chemically by the evolution of ammonia with alkali, but if this is missed they will be detected by their characteristic N—H *str* absorptions: also, they show two absorptions around 1650 cm^{-1} (the amide I and II bands; see charts, 3j). Secondary amides are always difficult to detect chemically; they are neutral and unreactive, but they show N—H *str* (see above) and two absorptions around 1600 cm^{-1}, only one of which may be unequivocally identified (the high-frequency band around 1680 cm^{-1}). Tertiary amides are less commonly encountered; they are also unreactive, show no N—H *str*, and only one band for C=O *str* around 1680 cm^{-1}.

If the compound contains nitrogen, absorbs around 1650–1750 cm^{-1}, and is not an amide, it may be an ester of (e.g.) an amino acid. The amine group should be easily detected, but a number of complications may arise, and these are discussed under amino acids (3n). Esters of *o*-amino acids have strong intramolecular hydrogen bonds (chelation) which lowers the C=O *str* frequency as shown on the charts under esters (2b).

Notes on the Correlation Charts

Units
Throughout this book, infrared data are quoted in wavenumbers (cm^{-1}), the units used by the majority of chemists. For the convenience of those chemists who prefer to use wavelength values (μm), the correlation charts show both wavenumbers and wavelengths.

Presentation
The correlation charts are presented in linear wavenumber form; but note the 'gear change' at 2000 cm^{-1}. Values above this figure are spaced more closely together than those below; this corresponds to the common 2:1 scale change which occurs at this point in many modern instruments.

In older infrared machines (for example those using a linear wavelength format) the region 3000–4000 cm^{-1} is particularly compressed, and this corresponds to relatively low resolution at these frequencies in the instrument. Provided the student remembers that the spectra only *appear* different, and that, whatever the instrument, the absorptions occur at the same wavenumber values, there should be no difficulty in correctly assigning the absorptions.

Note that the useful upper wavelength limit for sodium chloride optics is around 16 μm (625 cm^{-1}). Only a few of the more useful bands beyond that limit have been included on the charts (C—Cl, C—Br *str*, etc.).

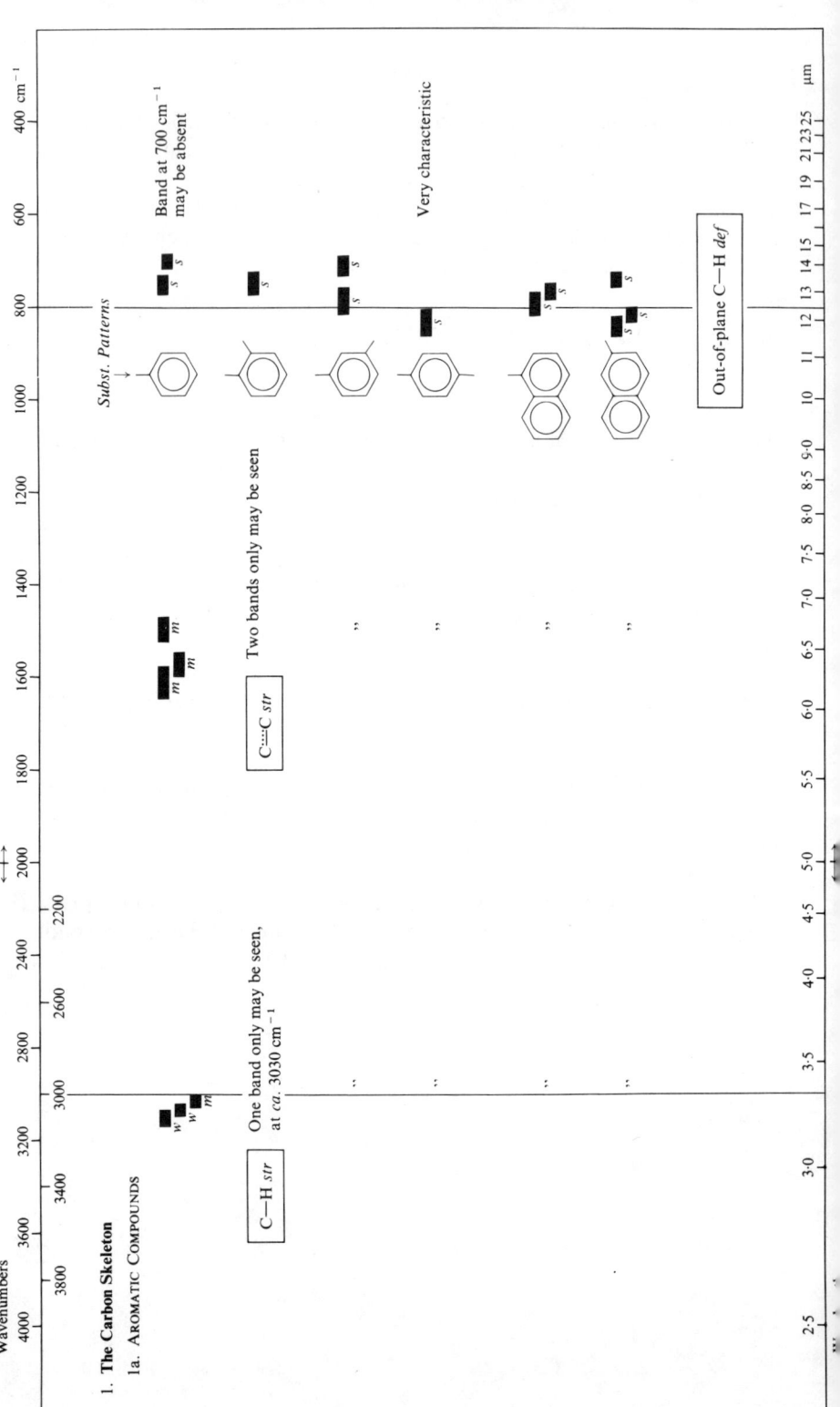

Wavenumbers

1. **The Carbon Skeleton**
1a. AROMATIC COMPOUNDS

Subst. Patterns →

Band at 700 cm⁻¹ may be absent

Very characteristic

Out-of-plane C—H *def*

Two bands only may be seen

C⃛C *str*

One band only may be seen, at *ca.* 3030 cm⁻¹

C—H *str*

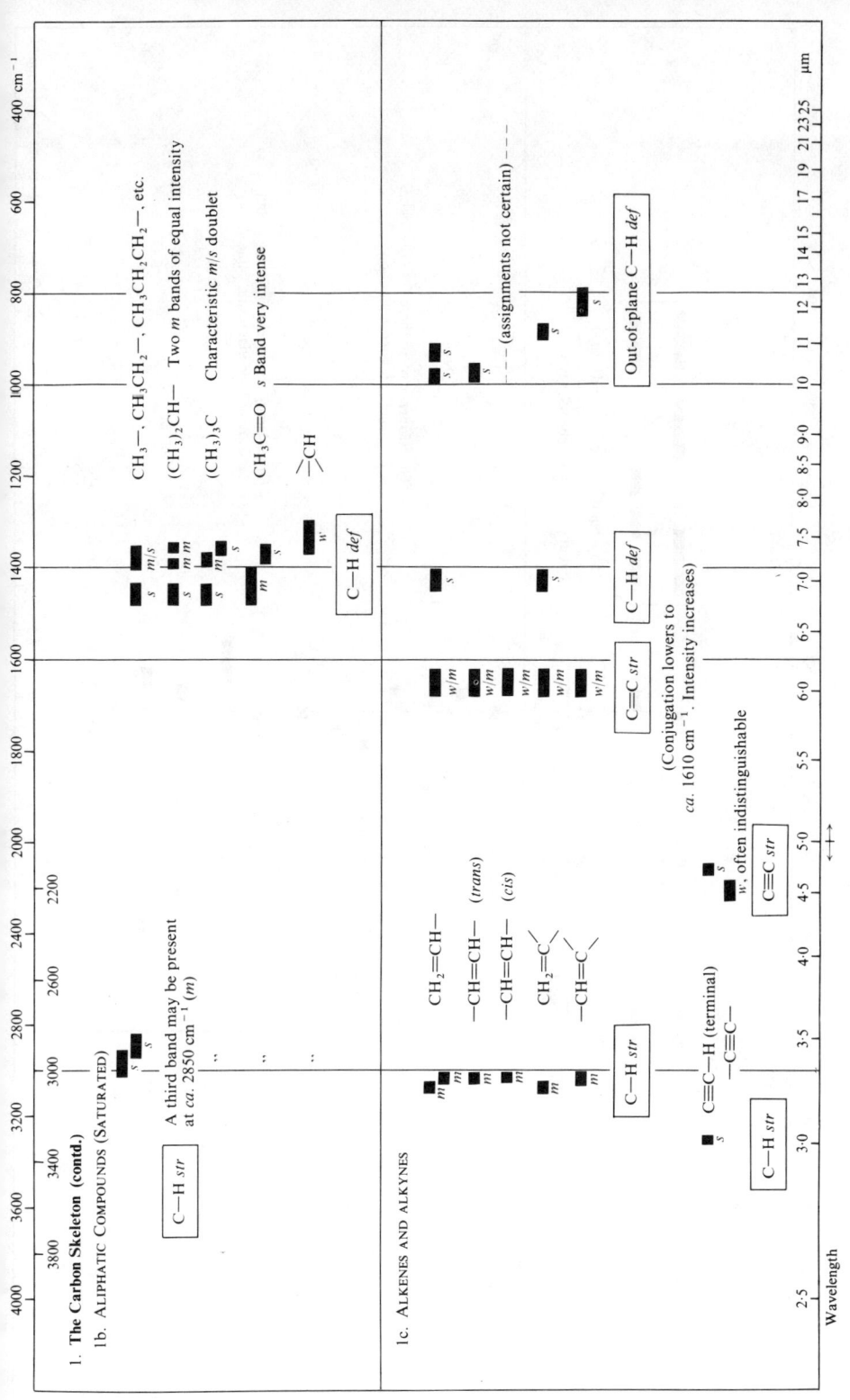

1. **The Carbon Skeleton (contd.)**

1b. ALIPHATIC COMPOUNDS (SATURATED)

A third band may be present at *ca.* 2850 cm^{-1} (*m*)

C—H *str*

C—H *str*

CH$_3$—, CH$_3$CH$_2$—, CH$_3$CH$_2$CH$_2$—, etc.

(CH$_3$)$_2$CH—   Two *m* bands of equal intensity

(CH$_3$)$_3$C   Characteristic *m/s* doublet

CH$_3$C=O   *s* Band very intense

$>$CH

C—H *def*

1c. ALKYNES AND ALKYNES

CH$_2$=CH—

—CH=CH— (*trans*)

—CH=CH— (*cis*)

CH$_2$=C$<$

—CH=C$<$

C—H *str*

C≡C—H (terminal)
—C≡C—

C≡C *str*

C=C *str*

(Conjugation lowers to *ca.* 1610 cm^{-1}. Intensity increases)

w, often indistinguishable

C—H *def*

Out-of-plane C—H *def*

—— (assignments not certain) ——

Wavelength

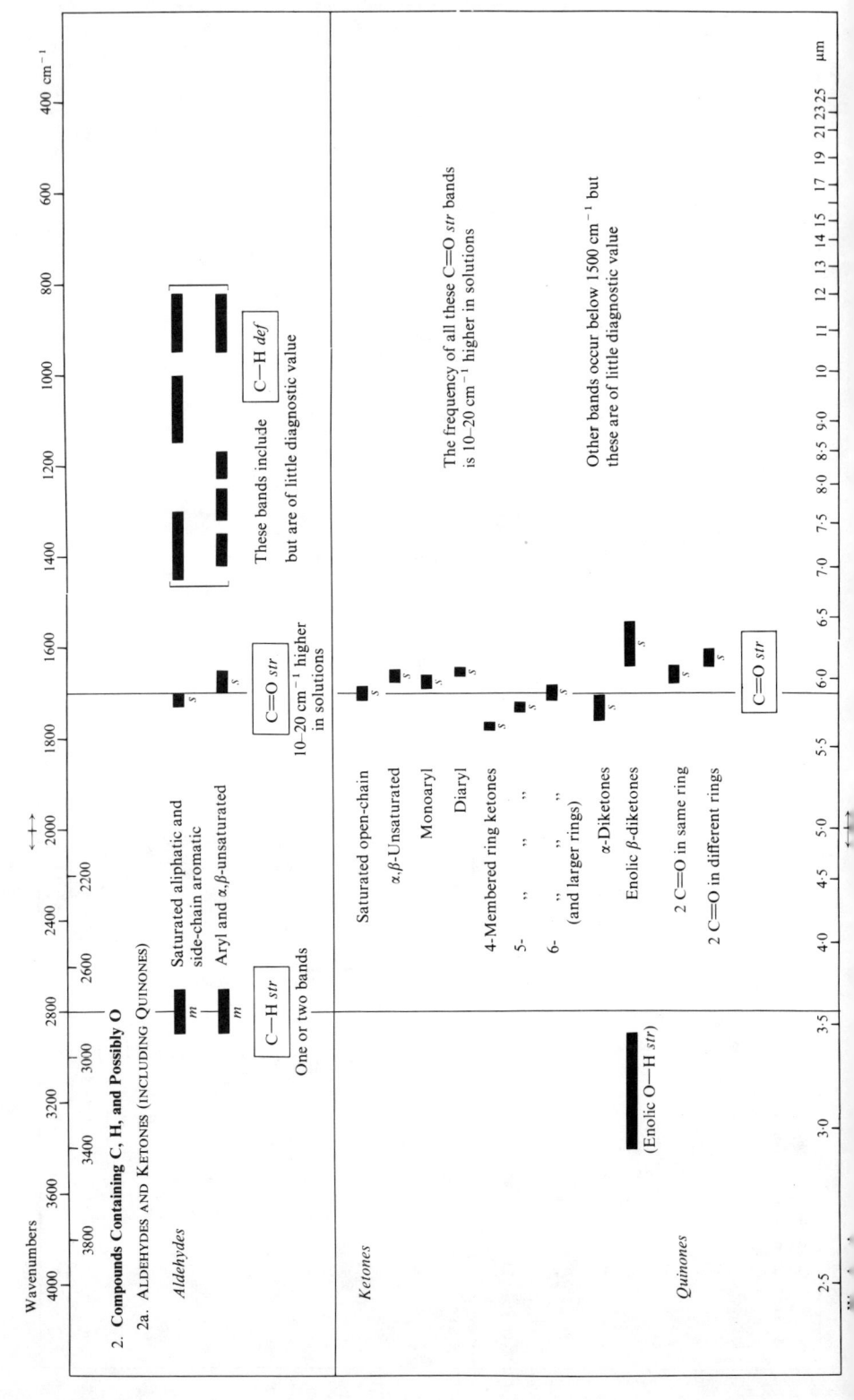

2. Compounds Containing C, H, and Possibly O

2a. Aldehydes and Ketones (including Quinones)

Wavenumbers

Aldehydes

Saturated aliphatic and side-chain aromatic

Aryl and α,β-unsaturated

C—H *str*
One or two bands

m

m

C=O *str*
10–20 cm⁻¹ higher in solutions

s

s

These bands include C—H *def*
but are of little diagnostic value

Ketones

Saturated open-chain

α,β-Unsaturated

Monoaryl

Diaryl

4-Membered ring ketones

5- ,, ,,

6- ,, ,, (and larger rings)

α-Diketones

Enolic β-diketones

2 C=O in same ring

2 C=O in different rings

(Enolic O—H *str*)

C=O *str*

The frequency of all these C=O *str* bands is 10–20 cm⁻¹ higher in solutions

Other bands occur below 1500 cm⁻¹ but these are of little diagnostic value

Quinones

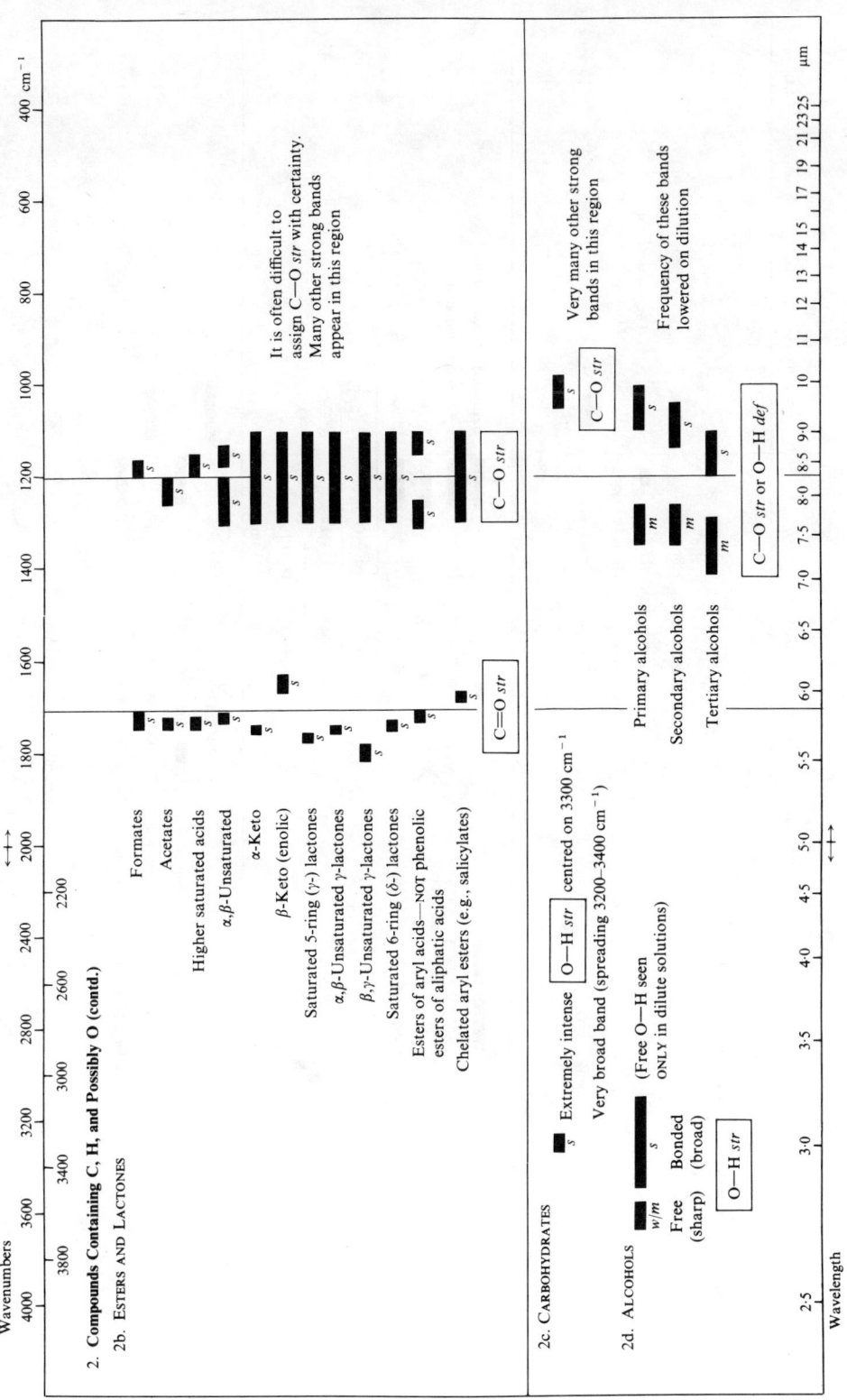

Wavenumbers

2. **Compounds Containing C, H, and Possibly O (contd.)**

2b. ESTERS AND LACTONES

It is often difficult to assign C—O *str* with certainty. Many other strong bands appear in this region

C—O *str*

C=O *str*

Formates

Acetates

Higher saturated acids

α,β-Unsaturated

α-Keto

β-Keto (enolic)

Saturated 5-ring (γ-) lactones

α,β-Unsaturated γ-lactones

β,γ-Unsaturated γ-lactones

Saturated 6-ring (δ-) lactones

Esters of aryl acids—NOT phenolic esters of aliphatic acids

Chelated aryl esters (e.g., salicylates)

2c. CARBOHYDRATES

Very many other strong bands in this region

C—O *str*

Frequency of these bands lowered on dilution

C—O *str* or O—H *def*

Extremely intense O—H *str* centred on 3300 cm⁻¹

Very broad band (spreading 3200–3400 cm⁻¹)

Primary alcohols

Secondary alcohols

Tertiary alcohols

2d. ALCOHOLS

(Free O—H seen ONLY in dilute solutions)

w/m Free (sharp)

s Bonded (broad)

O—H *str*

Wavelength

49

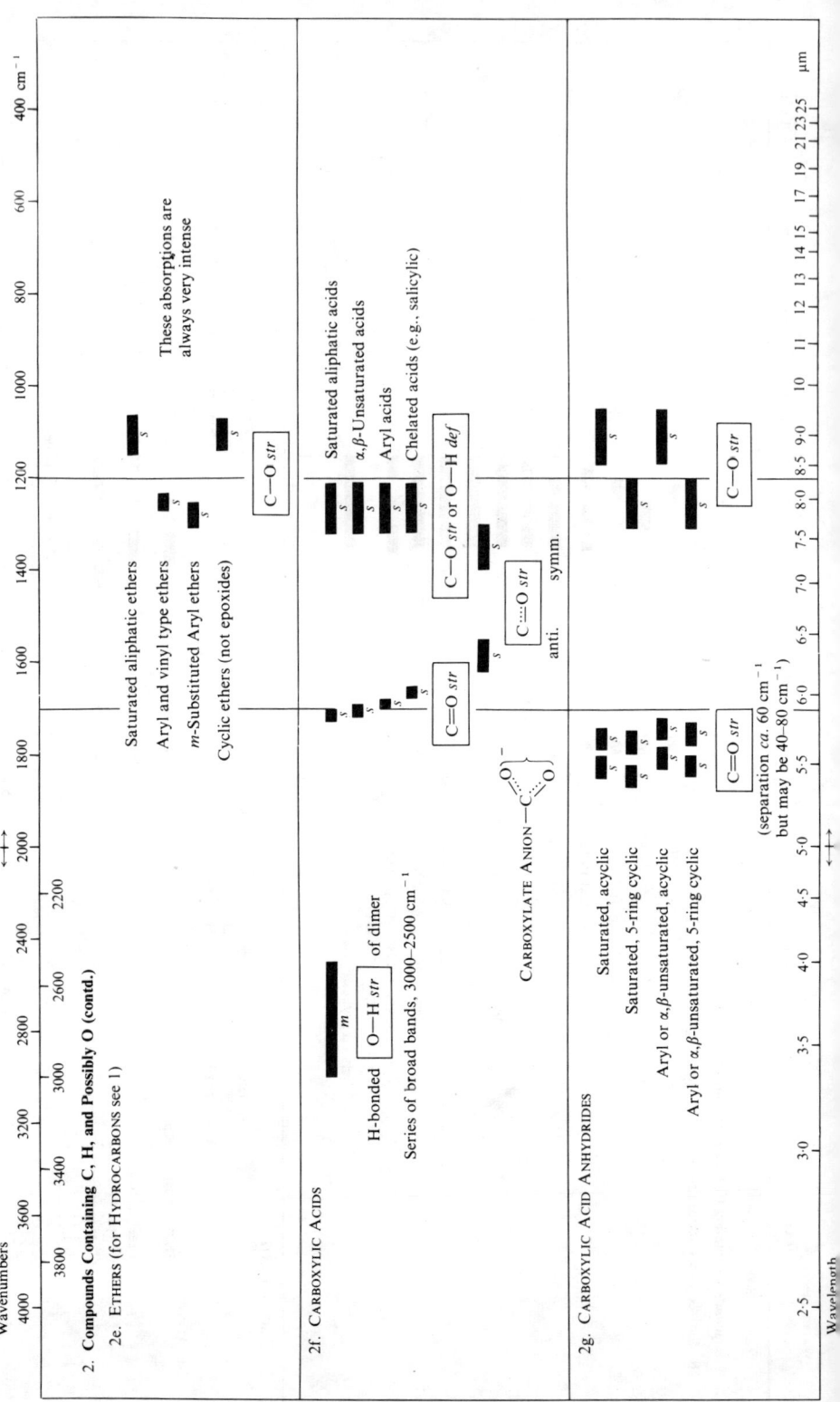

Wavenumbers

2. **Compounds Containing C, H, and Possibly O (contd.)**

2e. ETHERS (for HYDROCARBONS see 1)

Saturated aliphatic ethers

Aryl and vinyl type ethers

m-Substituted Aryl ethers

Cyclic ethers (not epoxides)

These absorptions are always very intense

C—O *str*

2f. CARBOXYLIC ACIDS

H-bonded | O—H *str* | of dimer

Series of broad bands, 3000–2500 cm⁻¹

CARBOXYLATE ANION

Saturated aliphatic acids

α,β-Unsaturated acids

Aryl acids

Chelated acids (e.g. salicylic)

C—O *str* or O—H *def*

C⋯O *str*   symm.

anti.

C=O *str*

2g. CARBOXYLIC ACID ANHYDRIDES

Saturated, acyclic

Saturated, 5-ring cyclic

Aryl or α,β-unsaturated, acyclic

Aryl or α,β-unsaturated, 5-ring cyclic

(separation *ca.* 60 cm⁻¹ but may be 40–80 cm⁻¹)

C=O *str*

C—O *str*

Wavelength

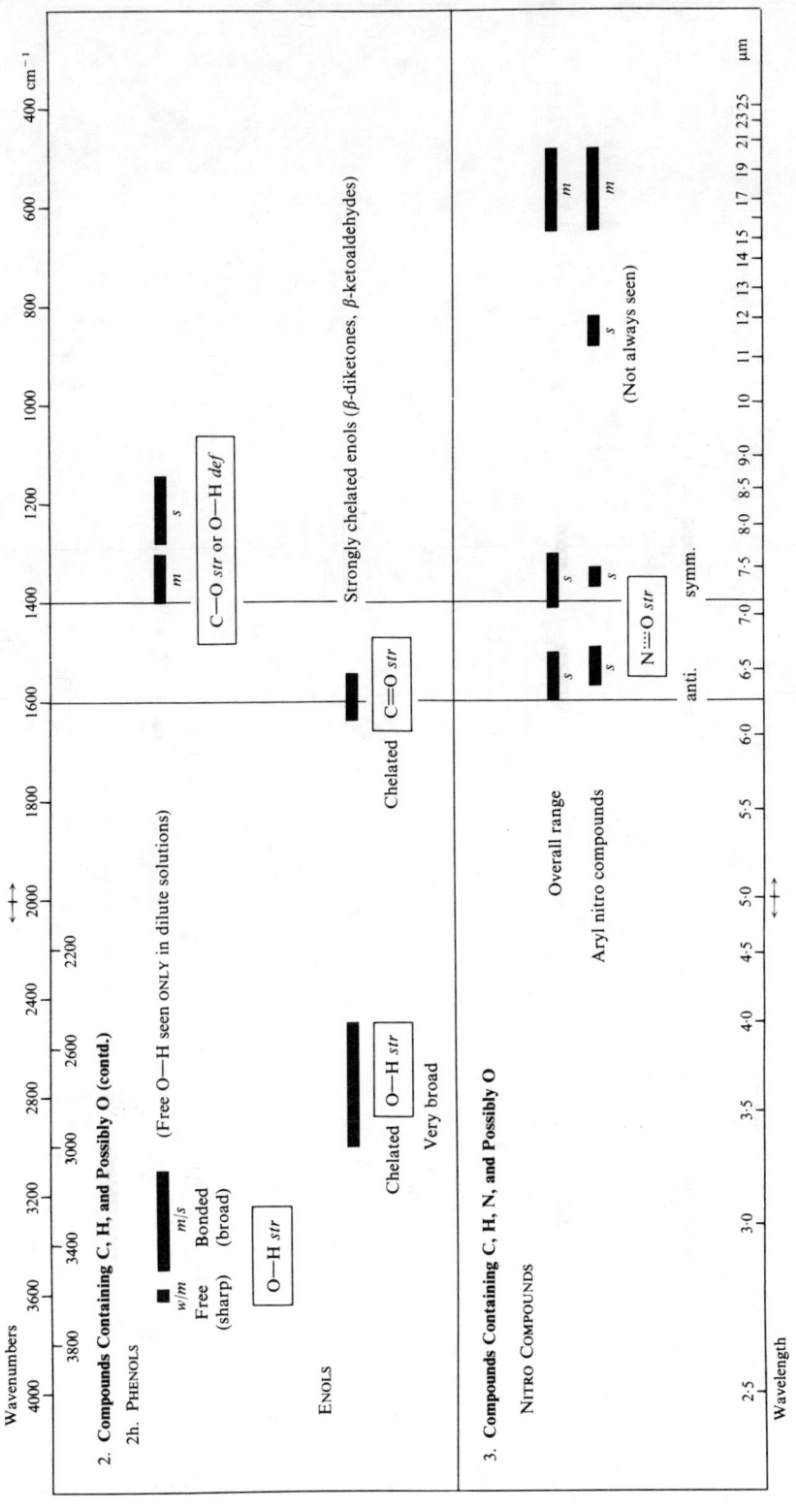

Wavenumbers

4000 3800 3600 3400 3200 3000 2800 2600 2400 2200 2000 1800 1600 1400 1200 1000 800 600 400 cm⁻¹

2. Compounds Containing C, H, and Possibly O (contd.)

2h. PHENOLS

(Free O—H seen ONLY in dilute solutions)

w/m   m/s
Free   Bonded
(sharp) (broad)

O—H str

m   s

C—O str or O—H def

ENOLS

Chelated

O—H str

Very broad

Chelated

C=O str

Strongly chelated enols (β-diketones, β-ketoaldehydes)

3. Compounds Containing C, H, N, and Possibly O

NITRO COMPOUNDS

Overall range

Aryl nitro compounds

s   s

s   s

s

m

m

(Not always seen)

N⋯O str

anti.   symm.

Wavelength

2·5 3·0 3·5 4·0 4·5 5·0 5·5 6·0 6·5 7·0 7·5 8·0 8·5 9·0 10 11 12 13 14 15 17 19 21 23 25 µm

51

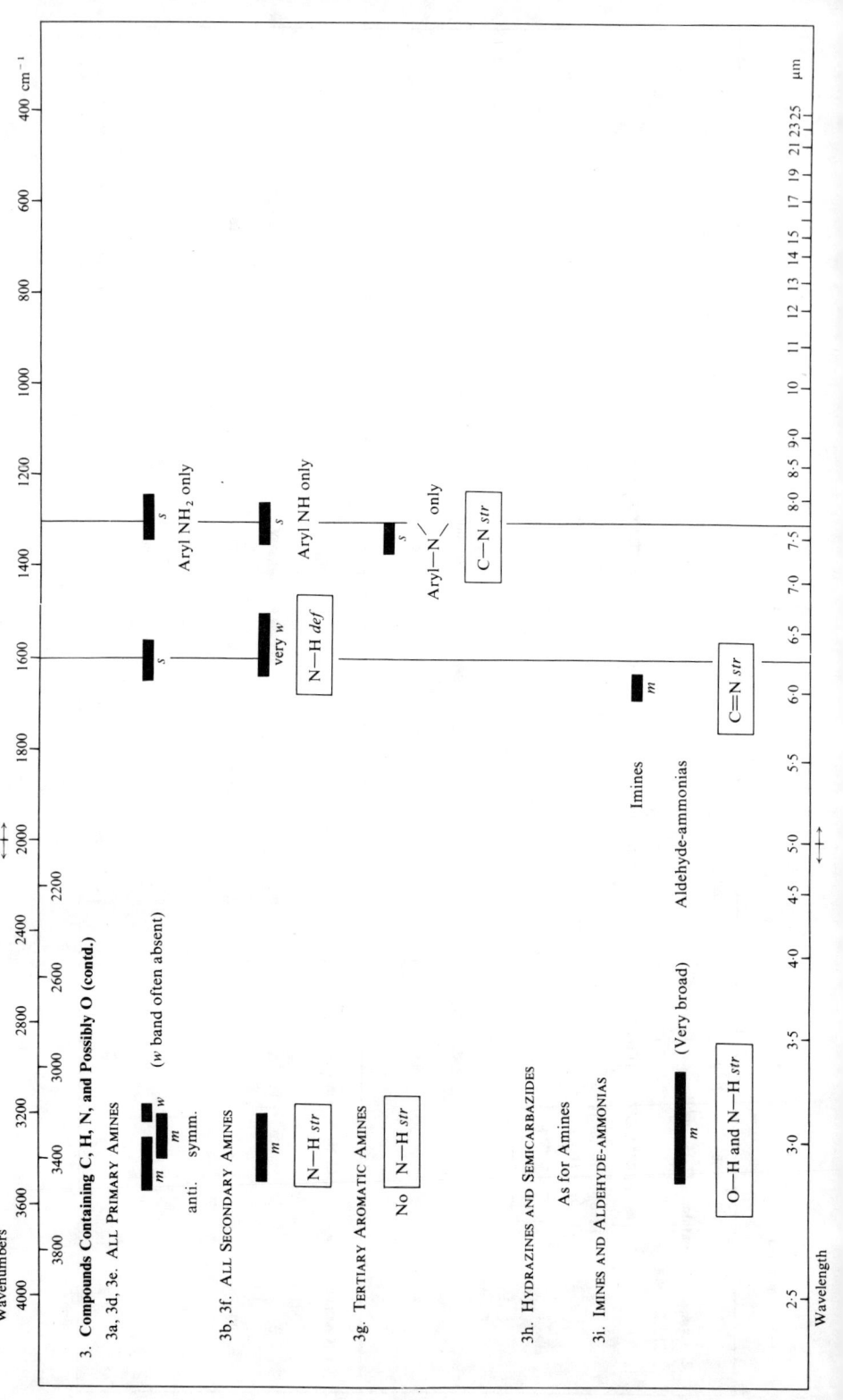

3. **Compounds Containing C, H, N, and Possibly O (contd.)**

3a, 3d, 3e. ALL PRIMARY AMINES

(*w* band often absent)

m ... *w*
anti. symm.

N—H *str*

Aryl NH₂ only | *s*
Aryl NH only | *s*
Aryl—N⟨ only | *s*
C—N *str*

N—H *def* | very *w* | *s*

3b, 3f. ALL SECONDARY AMINES

m

N—H *str*

3g. TERTIARY AROMATIC AMINES

No N—H *str*

C=N *str* | *m*

Imines

3h. HYDRAZINES AND SEMICARBAZIDES

As for Amines

Aldehyde-ammonias

3i. IMINES AND ALDEHYDE-AMMONIAS

(Very broad) | *m*

O—H and N—H *str*

Wavenumbers
4000 3800 3600 3400 3200 3000 2800 2600 2400 2200 2000 1800 1600 1400 1200 1000 800 600 400 cm⁻¹

Wavelength
2·5 3·0 3·5 4·0 4·5 5·0 5·5 6·0 6·5 7·0 7·5 8·0 8·5 9·0 10 11 12 13 14 15 17 19 21 23 25 µm

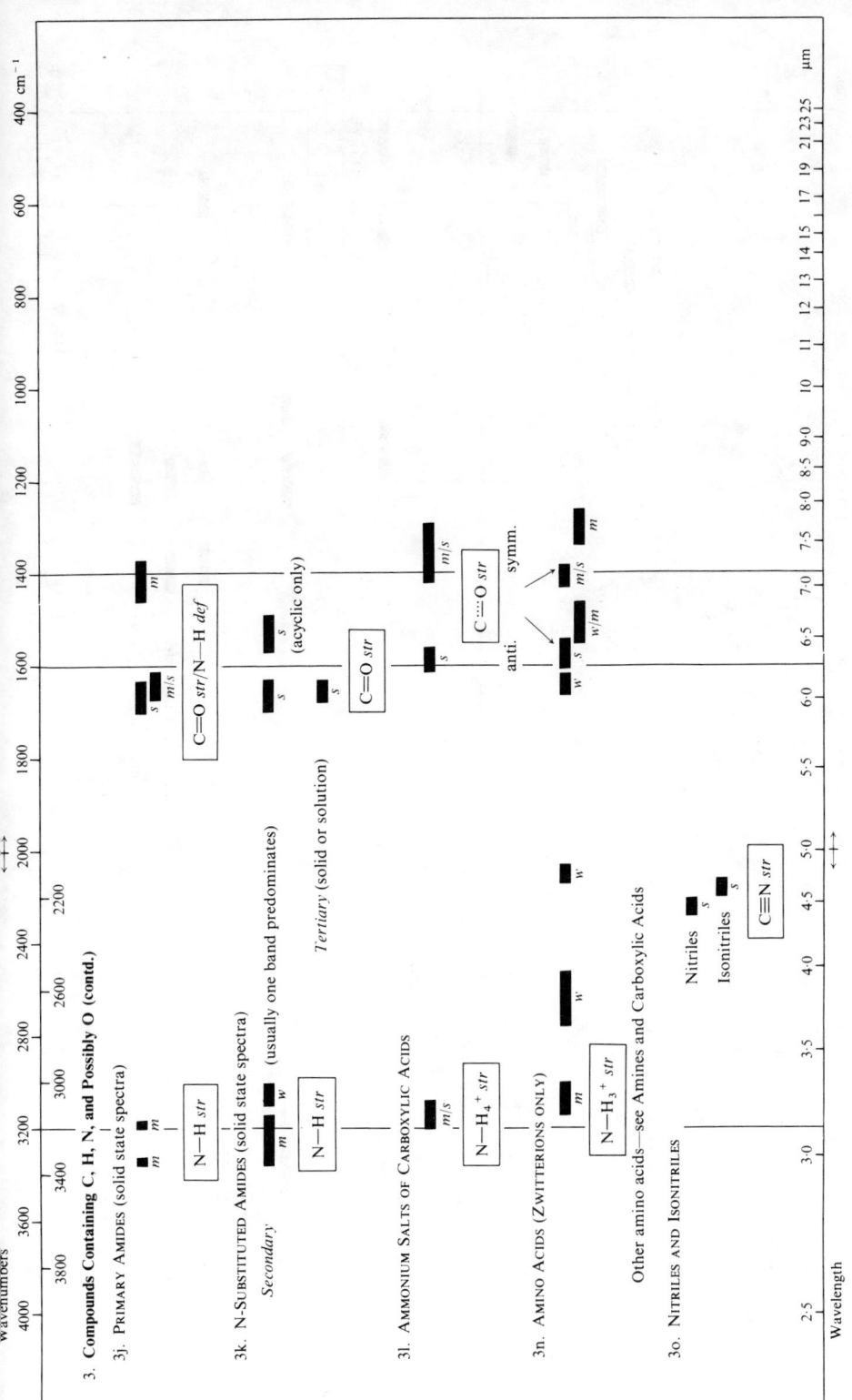

3. **Compounds Containing C, H, N, and Possibly O (contd.)**

3j. PRIMARY AMIDES (solid state spectra)

N—H str

C=O str/N—H def

3k. N-SUBSTITUTED AMIDES (solid state spectra)

Secondary (usually one band predominates)

N—H str

(acyclic only)

Tertiary (solid or solution)

C=O str

C⋯O str

symm.

anti.

3l. AMMONIUM SALTS OF CARBOXYLIC ACIDS

N—H₄⁺ str

3n. AMINO ACIDS (ZWITTERIONS ONLY)

N—H₃⁺ str

Other amino acids—see Amines and Carboxylic Acids

3o. NITRILES AND ISONITRILES

Nitriles

Isonitriles

C≡N str

Wavelength

53

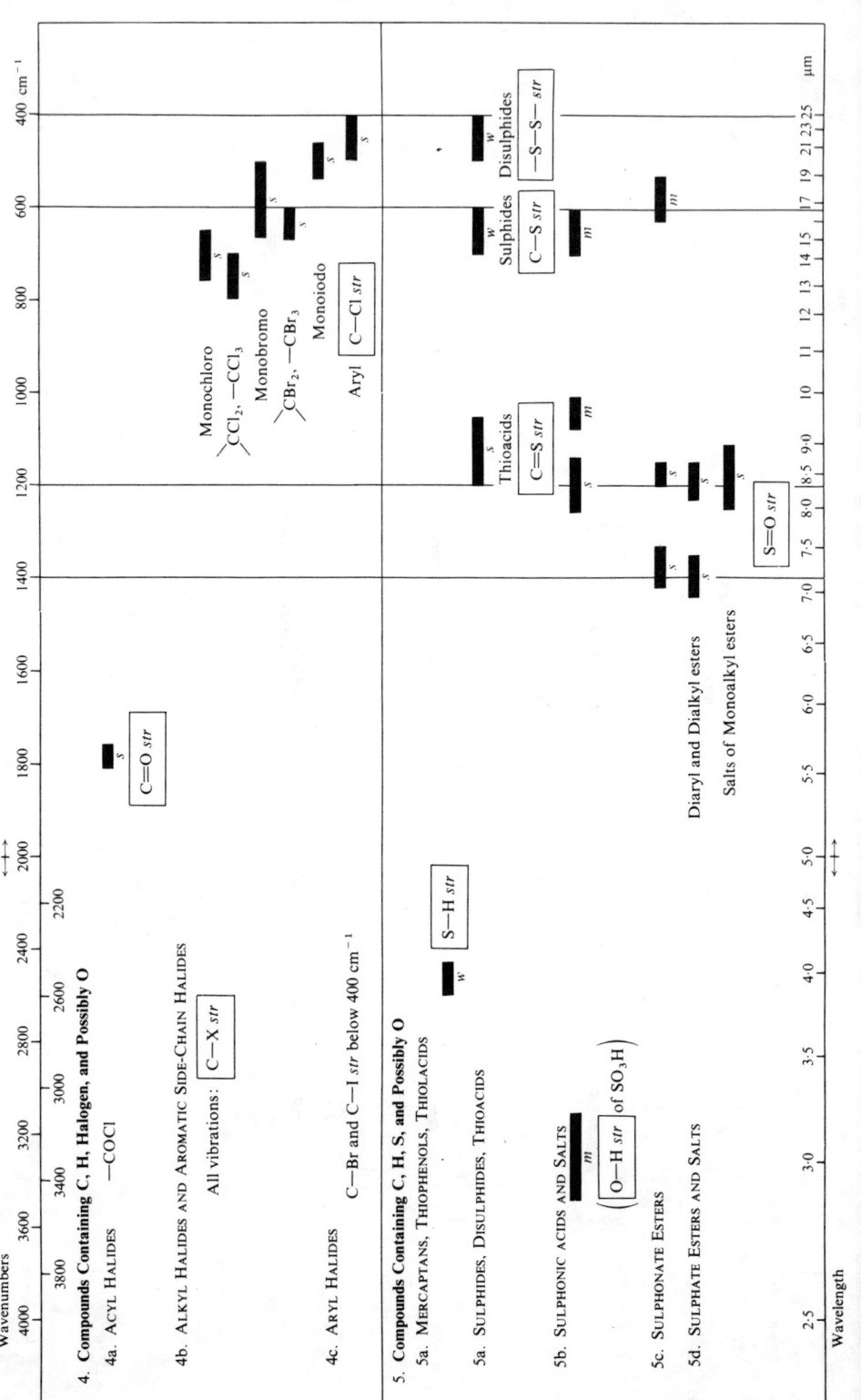

Wavenumbers

4. **Compounds Containing C, H, Halogen, and Possibly O**

4a. ACYL HALIDES   —COCl

$\boxed{C=O \text{ } str}$

4b. ALKYL HALIDES AND AROMATIC SIDE-CHAIN HALIDES

All vibrations: $\boxed{C-X \text{ } str}$

Monochloro
>CCl₂, —CCl₃

Monobromo
>CBr₂, —CBr₃

Monoiodo

Aryl $\boxed{C-Cl \text{ } str}$

4c. ARYL HALIDES   C—Br and C—I *str* below 400 cm⁻¹

5. **Compounds Containing C, H, S, and Possibly O**

5a. MERCAPTANS, THIOPHENOLS, THIOLACIDS

$\boxed{S-H \text{ } str}$

5a. SULPHIDES, DISULPHIDES, THIOACIDS

Disulphides $\boxed{-S-S- \text{ } str}$

Sulphides $\boxed{C-S \text{ } str}$

Thioacids $\boxed{C=S \text{ } str}$

5b. SULPHONIC ACIDS AND SALTS

$\boxed{\left(O-H \text{ } str\right) \text{ of } SO_3H}$

5c. SULPHONATE ESTERS

5d. SULPHATE ESTERS AND SALTS

Diaryl and Dialkyl esters

Salts of Monoalkyl esters

$\boxed{S=O \text{ } str}$

Wavelength

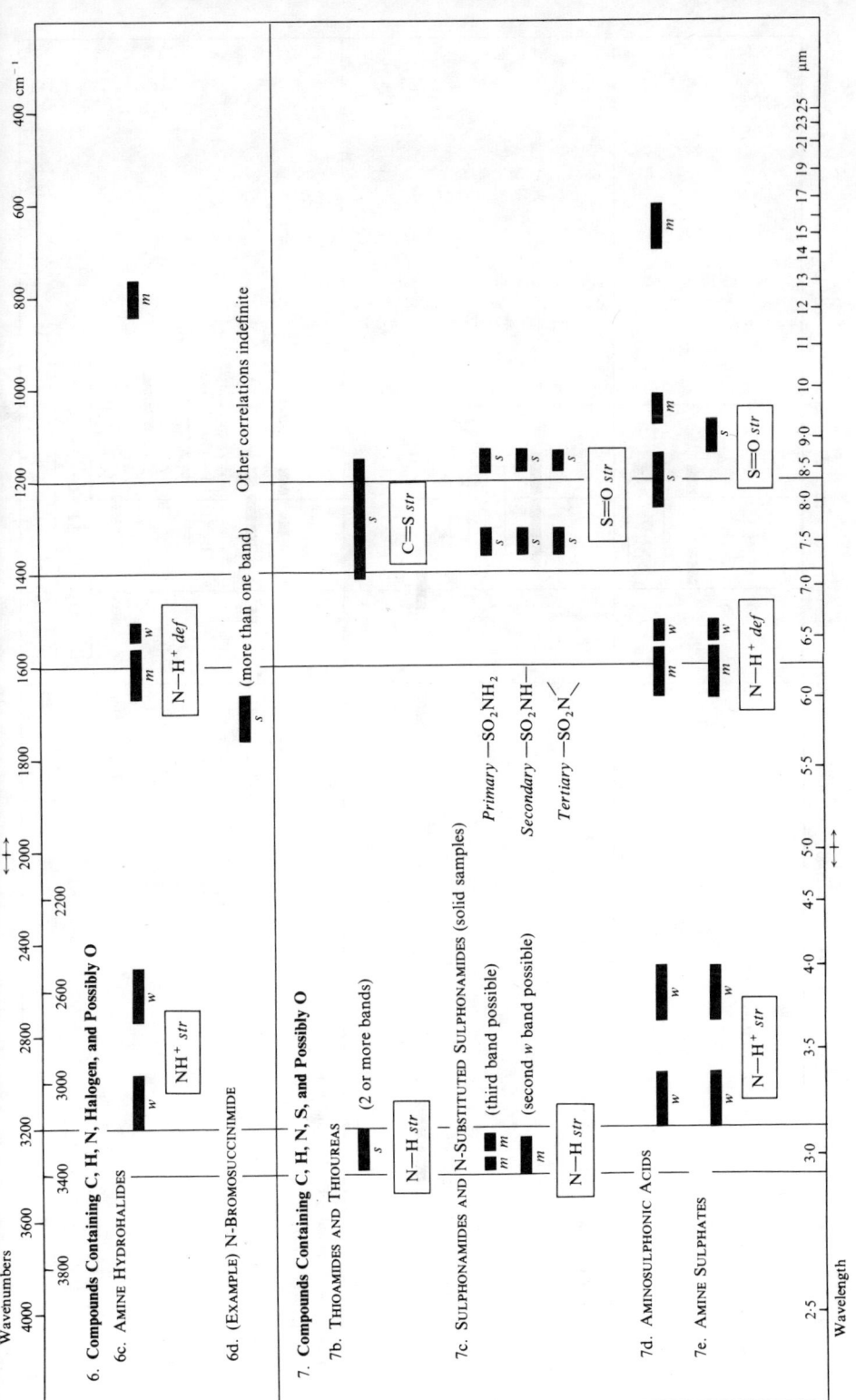

Wavenumbers

4000 3800 3600 3400 3200 3000 2800 2600 2400 2200 2000 1800 1600 1400 1200 1000 800 600 400 cm⁻¹

6. **Compounds Containing C, H, N, Halogen, and Possibly O**

6c. Amine Hydrohalides

NH⁺ *str*

N—H⁺ *def*

Other correlations indefinite

(more than one band)

6d. (Example) N-Bromosuccinimide

7. **Compounds Containing C, H, N, S, and Possibly O**

7b. Thioamides and Thioureas

N—H *str*

(2 or more bands)

C=S *str*

7c. Sulphonamides and N-Substituted Sulphonamides (solid samples)

N—H *str*

(third band possible)

(second *w* band possible)

Primary —SO₂NH₂

Secondary —SO₂NH—

Tertiary —SO₂N

S=O *str*

S=O *str*

7d. Aminosulphonic Acids

N—H⁺ *str*

N—H⁺ *def*

7e. Amine Sulphates

Wavelength

2·5 3·0 3·5 4·0 4·5 5·0 5·5 6·0 6·5 7·0 7·5 8·0 8·5 9·0 10 11 12 13 14 15 17 19 21 23 25 μm

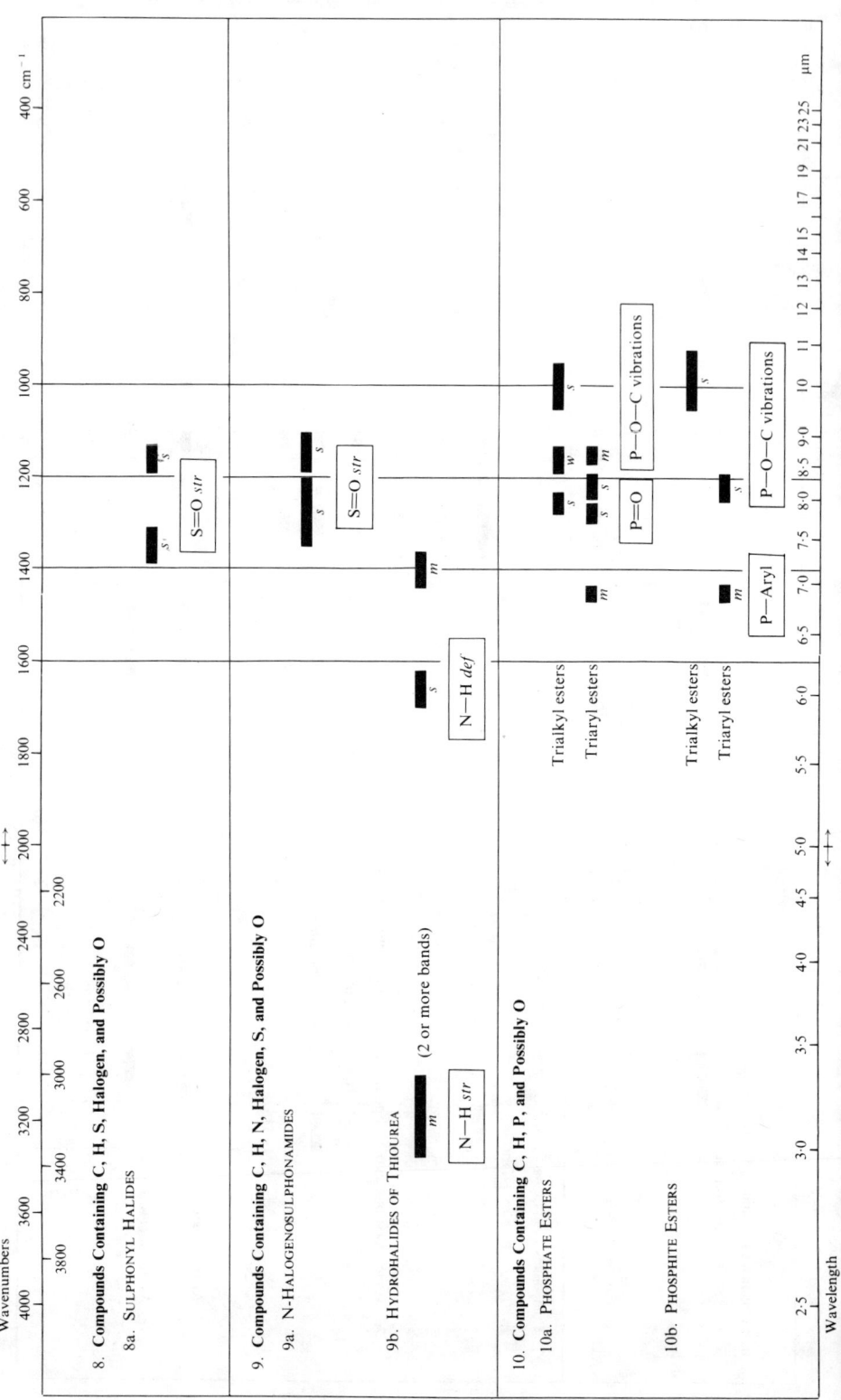

Wavenumbers

| 4000 | 3800 | 3600 | 3400 | 3200 | 3000 | 2800 | 2600 | 2400 | 2200 | 2000 | 1800 | 1600 | 1400 | 1200 | 1000 | 800 | 600 | 400 cm⁻¹ |

8. Compounds Containing C, H, S, Halogen, and Possibly O

8a. SULPHONYL HALIDES

S=O *str*

9. Compounds Containing C, H, N, Halogen, S, and Possibly O

9a. N-HALOGENOSULPHONAMIDES

S=O *str*

9b. HYDROHALIDES OF THIOUREA

(2 or more bands)

N—H *str*

N—H *def*

10. Compounds Containing C, H, P, and Possibly O

10a. PHOSPHATE ESTERS

Trialkyl esters

Triaryl esters

P=O

P—O—C vibrations

10b. PHOSPHITE ESTERS

Trialkyl esters

Triaryl esters

P—Aryl

P—O—C vibrations

Wavelength

| 2·5 | 3·0 | 3·5 | 4·0 | 4·5 | 5·0 | 5·5 | 6·0 | 6·5 | 7·0 | 7·5 | 8·0 | 8·5 9·0 | 10 | 11 | 12 | 13 14 15 | 17 | 19 | 21 23 25 μm |

4

Identification, Derivatives, Tables of Physical Constants

4.1 Introduction

Methods are given here for the preparation of a wide range of derivatives of the functional classes met in chapter 2. Where a convenient quantitative method is available for measuring the equivalent weight of the compound, this is also given; since this gives considerable help in identifying the compound, it should be carried out wherever possible, although it does not distinguish among isomers.

Before preparing a derivative, refer to the tables at the end of this chapter and tentatively identify the compound from its m.p. or b.p. or other known property. If two or more possibilities exist on the basis of this information, then *choose a derivative whose m.p. will clearly distinguish the various possible structures.*

For example, butan-1-ol and pentan-2-ol have similar boiling points (118° and 119° respectively). The 3,5-dinitrobenzoate esters would not distinguish these alcohols, since the melting points are again close (64° and 62° respectively); it would therefore be fruitless to prepare this derivative. The 1-naphthylurethane derivatives would be little better (melting points 72° and 76° respectively). Differentiation could only be achieved by preparing the hydrogen 3-nitrophthalate derivative, since the melting points are substantially different (147° and 103° respectively).

4.2a Aldehydes and Ketones (and Quinones)

2,4-Dinitrophenylhydrazones

In a dry flask place 2,4-dinitrophenylhydrazine (0·4 g), and methanol or ethanol (10 cm³): slowly pour in concentrated sulphuric acid (0·5 cm³) and warm until a clear solution is obtained. While the solution is still hot, add the unknown (0·3 g): if no precipitate appears immediately, warm the solution for 2 minutes and cool. A *few* drops of water may also be added.

Filter off and recrystallize from ethanol, ethyl acetate, or chloroform. Recrystallization may be difficult, and is not always necessary.

$$-\overset{\displaystyle O}{\underset{\displaystyle \|}{C}}\;\overset{\displaystyle H^+}{\nearrow} \longrightarrow \;-\overset{\displaystyle OH}{\underset{\displaystyle |}{C^+}}\; \overset{\frown}{NH_2.NH}\!-\!\!\bigcirc\!\!-\!\!NO_2 \overset{-H_2O}{\longrightarrow}$$

(with NO_2 at top of ring)

$$\overset{\displaystyle NO_2}{} \\ \overset{\displaystyle >}{}C\!=\!N.NH\!-\!\!\bigcirc\!\!-\!\!NO_2 + H^+$$

For enolic β-diketones, the condensation may give a pyrazole derivative (a phenylpyrazolone).

p-Nitrophenylhydrazones
Reflux briefly together the unknown (0·4 g), p-nitrophenylhydrazine (0·4 g), ethanol (10 cm^3), and glacial acetic acid (0·2 cm^3). While still warm add a few drops of water until the solution *just* becomes cloudy: cool, filter off, and recrystallize from ethanol.

Equation and further notes are as for dinitrophenylhydrazones.

Semicarbazones
In a small flask prepare a solution of semicarbazide hydrochloride (0·5 g), and sodium acetate dihydrate (0·8 g) in water (5 cm^3); add the unknown (0·5 g) and warm on a water bath for 10 minutes. Cool in ice; if no precipitation occurs, return the flask to the water bath for a further 10 minutes and again cool in ice. (Longer reaction times may be needed.) Filter off the precipitated derivative, wash with ice-water, and recrystallize from ethanol.

$$-\overset{\displaystyle O}{\underset{\displaystyle \|}{C}}\;\overset{\displaystyle H^+}{\nearrow} \longrightarrow \;-\overset{\displaystyle OH}{\underset{\displaystyle |}{C^+}}\; \overset{\frown}{NH_2.NH.CO.NH_2} \overset{-H_2O}{\longrightarrow} \;{>}C\!=\!N.NH.CO.NH_2 + H^+$$

Oximes (not recommended for Aliphatic Aldehydes)
For water-soluble compounds use the method given for semicarbazides, but substituting hydroxyammonium chloride (hydroxylamine hydrochloride) for the semicarbazide hydrochloride.

For others, use as solvent the minimum quantity of aqueous methanol or ethanol to ensure a homogeneous solution; cooling in ice/salt may be necessary to precipitate the oxime.

Recrystallize the oxime from water or the *minimum* of aqueous ethanol.

$$-\overset{\displaystyle O}{\underset{\displaystyle \|}{C}}\;\overset{\displaystyle H^+}{\nearrow} \longrightarrow \;-\overset{\displaystyle OH}{\underset{\displaystyle |}{C^+}}\; \overset{\frown}{NH_2OH} \overset{-H_2O}{\longrightarrow} \;{>}C\!=\!N.OH + H^+$$

Dimedones (Aldehydes only)

Dissolve dimedone (0·5 g) in aqueous ethanol (1:1, 5–10 cm³), and dissolve the aldehyde (0·4 g) separately in the same solvent. Mix the solutions and add a drop of piperidine; the derivative crystallizes on standing. If no precipitation occurs, warm the reaction mixture gently for a few minutes and again cool; continue thus until precipitation occurs. Filter off and recrystallize from *aqueous* alcohol.

dimedone   aldehyde dimedone

An infrared study of this derivative gives interesting information on the enolic nature of the derivative.

Dimedone Anhydrides

Dissolve the aldehyde dimedone derivative (0·2 g) in 80 per cent aqueous ethanol (5–10 cm³), add a drop of concentrated hydrochloric acid, and reflux for 5 minutes. While still warm, add drops of water until the solution just remains cloudy then cool in ice. Filter off and recrystallize from aqueous ethanol.

dimedone anhydride

Hydroquinone Acetate (Quinones only)

To a small dry flask add acetic anhydride (3 cm³) and the unknown quinone (0·5 g); add zinc powder (0·5 g) and *anhydrous (fused)* sodium acetate (0·2 g). Warm the mixture gently, then finally reflux for 5 minutes. Add glacial acetic acid (3 cm³) and again heat to boiling. Decant off the solution and warm it gently while adding water to hydrolyse the acetic anhydride; to reduce the bulk, dilute alkali may be added dropwise to complete the removal of anhydride.

When the smell of acetic anhydride has disappeared, cool in ice/salt. Filter off and recrystallize from aqueous ethanol.

e.g.,

4.2b   Esters (and Lactones)

Equivalent Weight

In a 250 cm^3 flask, dissolve *ca.* 3 g of potassium hydroxide in diethylene glycol (50 cm^3) containing a small amount (2–3 cm^3) of water. Heat to aid dissolution, then cool and add a further 50 cm^3 of diethylene glycol. The resulting solution is *ca.* 0·5M.

Pipette 25 cm^3 of this solution into a 50 cm^3 flask with standard taper joint; use a safety pipette with a hand aspirator. Now add the ester (*ca.* 1 g, accurately weighed) and reflux for 10 minutes. Cool and wash down the condenser with distilled water, then transfer the solution with washing to a conical flask.

Titrate with standard hydrochloric acid (*ca.* 0·2M) using phenolphthalein as indicator.

Run a simultaneous duplicate and a blank estimation.

From these figures, calculate the equivalent weight of the ester.

The conditions given above are suitable for almost all esters; those of a very high molecular weight, such as glycerides, may require longer periods of refluxing.

Hydrolysis to the Acid and Alcohol, or Phenol

The surest identification of an ester lies in identifying the constituent acid and alcohol; a number of different isolation procedures must be used, depending on the volatility and solubility of the acid and alcohol. Choose the most appropriate method from among the following.

(i) **Simple Aliphatic Esters** (2–3 g) can be hydrolysed by 0·5–1 hour reflux with 20 per cent aqueous sodium hydroxide (25 cm^3): the mixture may become homogeneous as the reaction proceeds, unless the alcohol is C$_5$ or higher. After the reflux try to distil (or steam distil) the free alcohol from the alkaline solution: it may be necessary to extract the alcohol with ether, either from the distillate (if it is steam volatile) or from the alkali residue (if it does not distil over). When the alkali residue is acidified (concentrated hydrochloric acid) the free acid will be liberated: it may crystallize out on cooling, but may have to be extracted with ether.

Having isolated the alcohol and acid, identify them individually (class 2d and class 2f respectively).

This procedure will not enable you to isolate the acid if it happens to be water soluble, but is insoluble in ether. In such cases a derivative can be made directly on the solution of the salt without isolating the free acid. (See for example the preparation of S-benzylthiuronium derivatives of acids.)

(ii) **Aromatic esters** can often be treated as in (i), while others may require a 1 hour reflux with 10 per cent ethanolic potassium hydroxide to effect hydrolysis. In this case, since ethanol is used as a solvent, it is not practicable to try and isolate lower aliphatic alcohols from the ester, and identification must rest on characterization of the acid, etc. Before acidification to liberate the free acid, distil off most of the ethanol; thereafter proceed as in (i).

(iii) **Phenolic esters** on hydrolysis give an acid and a phenol, both of which form salts in the alkali; phenols can be steam-distilled out or extracted only after the solution is made acid by passing in CO_2. Identify the phenol (class 2h).

(iv) **Lactones** hydrolyse to the salt of the hydroxy acid in alkali, but on acidification usually revert to the lactone form spontaneously. Identification is normally possible on the basis of equivalent weight.

(v) **Difficult hydrolyses**: esters which are not hydrolysed by aqueous or ethanolic alkali should be refluxed with alkali in diethylene glycol. Add the ester (2–3 g) to a solution of potassium hydroxide (1 g) in diethylene glycol (5 cm^3) and water (0·2 cm^3). Reflux the mixture for 10 minutes, cool, and proceed as in (i), (ii), or (iii) as appropriate.

4.2c   Carbohydrates

Specific Rotation
Values given in table 4.8 refer, where appropriate, to the mutarotated mixture of anomeric forms. Add a few drops of aqueous ammonia to ensure rapid mutarotation.

Acetates
(i) The α-*Acetate* is obtained using acid catalysis (zinc chloride): in a 50 cm^3 round bottom flask place acetic anhydride (10 cm^3) and powdered anhydrous (granular) zinc chloride (0·5 g). Heat under reflux for 5 minutes on a water bath, then add the carbohydrate (0·3 g) and heat for a further hour. Pour with stirring into 100 cm^3 of ice-water and stir with cooling until the acetate crystallizes. Filter off and recrystallize from ethanol.

(ii) The β-*Acetate* is obtained using base catalysis (sodium acetate):

repeat the above procedure, substituting *anhydrous* sodium acetate (1 g) for zinc chloride: heat for a total of 2 hours.

a-D-(+)-glucose   the a-acetate

Phenylosazones
The melting points and physical appearance of osazones do not lead to reliable identification of carbohydrates, and have been omitted.

Help in identification can be obtained by noting the time taken for the osazone to form under standard conditions; in the case of sucrose, the osazone formed is glucosazone, from the glucose and fructose produced on hydrolysis. Into a clean test-tube pipette 5 cm^3 of water; to this add the sugar (0·25 g), crystalline sodium acetate (0·75 g), and analytical grade phenylhydrazine hydrochloride (0·5 g). Stand the tube in boiling water and note the time taken for the osazone to form.

Chromatography
Where authentic samples of sugars are available, paper chromatography or thin layer chromatography (T.L.C.) on silica gel can be used to identify them. The unknown sugar is spotted onto the paper or plate together with known sugars (5 μl of 10 per cent solutions), and the chromatogram developed with propan-1-ol: water (78:22). Visualization can be with aniline hydrogen oxalate spray (2 per cent in ethanol), or, in the case of T.L.C., with 50 per cent sulphuric acid followed by heating to 150°.

Comparison of the distance moved by the unknown sugar with the distance moved by the known sugars will establish the identity of the unknown.

Gas chromatography of trimethylsilyl ethers is outside the scope of this book.

4.2d   Alcohols

3,5-Dinitrobenzoates
(i) **Preparation of 3,5-dinitrobenzoyl chloride** is carried out as follows: equal quantities (*ca.* 1 g) of 3,5-dinitrobenzoic acid and phosphorus pentachloride are mixed together in a dry test-tube. On warming, vigorous evolution of HCl occurs—*carry out this stage in the fume cupboard*. Stir with a glass rod until the reaction subsides, then continue fairly steady heating for a further 3–4 minutes, and cool the test-tube in cold water: the

mixture becomes semi-solid. Transfer the mass to a piece of unglazed porcelain, and work the paste about with a spatula until the phosphoryl chloride is fully absorbed. The remaining crude acid chloride is adequately pure for the next stage.

(ii) **Preparation of the 3,5-dinitrobenzoate ester** is carried out by heating together the alcohol and an equimolar amount of 3,5-dinitrobenzoyl chloride. For the lower aliphatic alcohols, a slight excess of alcohol is not serious, but for oily alcohols *this must be avoided* or difficulty will be experienced in crystallizing the derivative. Mix the two reactants in a test-tube, while heating steadily in a boiling water bath; evolution of HCl should stop after 3–5 minutes. (As an alternative, add 2 ml of pyridine to the reaction and reflux the mixture, in which case no HCl will be evolved. Pour into water, filter off the residue, and continue as follows.) Cool, and wash the residue with aqueous sodium bicarbonate to remove 3,5-dinitrobenzoic acid (m.p. 207°), and recrystallize from ethanol, chloroform, or benzene.

p-Nitrobenzoates

These are prepared as for 3,5-dinitrobenzoates, preferably using the pyridine modification.

1-Naphthylurethanes

The reagent, 1-naphthyl isocyanate, must be protected from water, and the alcohol must be dry: for the lower aliphatic alcohols and liquid dihydric alcohols, which may contain substantial amounts of water, treat them for a few minutes with a small amount of anhydrous magnesium sulphate, before decanting off.

In a dry test-tube place the alcohol (1 g) and 1-naphthyl isocyanate (0·5 cm³); heat in an oil bath at 100° (*or lightly* cork the tube to exclude moisture and heat in a water bath) for 5–10 minutes then cool in ice. Stir the mixture with a glass rod until crystallization occurs, then recrystallize from light petroleum (b.p. 60–80°) or carbon tetrachloride. If the hot recrystallizing solution contains a residue, it is likely to be N,N'-bis-(1-naphthyl)-urea, formed when the reagent reacts with water: it should be filtered off and discarded.

a carbamate ester
(or urethane)

Hydrogen 3-Nitrophthalates

(i) **3-Nitrophthalic anhydride**: reflux 3-nitrophthalic acid (1 g) with acetic anhydride (2 cm^3) for 15 minutes then cool in ice/salt. Filter off the 3-nitrophthalic anhydride and wash with a little ether. Protect from moisture: the product is sufficiently pure for the next stage.

(ii) **Hydrogen 3-nitrophthalate** esters are formed quickly from lower aliphatic alcohols: heat the reagent (1 g) with the alcohol (0·5 g) on a water bath for 30 minutes, then recrystallize the product from water. For the higher alcohols, reflux together the reagent (1 g), the alcohol (0·5 g) and toluene (5 cm^3) for 2–3 hours, or until the solution becomes homogeneous, then cool in ice/salt. Filter off the product and recrystallize from aqueous ethanol.

Equivalent Weight of Hydrogen 3-Nitrophthalates

Since these esters have a free carboxyl group, their equivalent weights can be measured by alkali titration: see 4.2f.

4.2e   Ethers and Hydrocarbons

3,5-Dinitrobenzoates (Symmetrical Aliphatic Ethers)

Symmetrical aliphatic ethers react with 3,5-dinitrobenzoyl chloride to give the 3,5-dinitrobenzoate ester of the corresponding alcohol.

This reaction cannot be used for unsymmetrical ethers, which are best identified from b.p. data.

Reflux together for 1 hour the ether (1 g), 3,5-dinitrobenzoyl chloride (0·5 g) and anhydrous (granular) zinc chloride (0·1 g): cool, and continue the work up as described for alcohols (4.2d) starting from the treatment with sodium bicarbonate.

Consult table 4.9 for the m.p. of these esters.

Picrates (Aromatic Ethers and Hydrocarbons)
Picrates of monocyclic aromatic hydrocarbons are in general only formed
in solution, and cannot be used as derivatives: consult table 4.16 before
preparing the picrate of a hydrocarbon.

Dissolve the unknown hydrocarbon or ether (0·5 g) in the minimum of
hot ethanol or benzene. Dissolve the same amount of picric acid in the
minimum of the same solvent, also hot. Mix the two hot solutions; the
charge transfer complex (picrate) forms immediately. Cool, and filter off.
Wash with a small amount of ethanol, but stop the washing if the picrate
changes colour (i.e., decomposes to its constituents). Do not attempt to
recrystallize hydrocarbon picrates, since they frequently decompose;
prepare the picrate with pure solutions and it will usually crystallize out
from the reaction in a pure state.

e.g. anthracene

Aroylbenzoic Acids
Friedel–Crafts acylation of an aromatic hydrocarbon with phthalic
anhydride gives the o-aroylbenzoic acid. A study of the infrared spectra of
these derivatives is interesting, since they do not exist as keto acids, but as
lactols (see below).

The equivalent weight of the acids can also be determined, as described
in 4.2f.

In a 50 cm^3 round bottom flask place methylene chloride (10 cm^3) and
the unknown hydrocarbon (1 g): add phthalic anhydride (1·2 g) and *freshly*
powdered anhydrous aluminium chloride (2·5 g). Reflux on a boiling
water bath for 30 minutes, then pour on to ice (10 g) and concentrated
hydrochloric acid (10 cm^3). Transfer the whole to a separating funnel, and
shake intermittently until all the aluminium salts dissolve: discard the
lower aqueous layer. Carefully add dilute sodium carbonate solution and
shake the funnel, releasing CO_2 frequently, until no more CO_2 is produced.
Draw off the carbonate layer, which contains the sodium salt of the deriva-

tive, and slowly acidify with concentrated hydrochloric acid. Filter off the free o-aroylbenzoic acid and recrystallize from aqueous ethanol.

the aroylbenzoic acid   the lactol

Nitro Derivatives (Aromatic Ethers and Hydrocarbons)

WARNING: DO NOT BEGIN A NITRATION UNLESS YOU ARE ABSOLUTELY CERTAIN THAT YOUR UNKNOWN IS AN AROMATIC ETHER OR HYDROCARBON. CARRY OUT THE EXPERIMENT IN THE FUME CUPBOARD, AND INFORM YOUR INSTRUCTOR BEFORE YOU BEGIN.

It is not possible to predict the conditions necessary to nitrate an unknown compound. Following this are three different sets of conditions, and you are advised to try method (i) first; if this does not produce a solid nitro derivative of your unknown, try (ii), and then (iii).

(i) Slowly add concentrated sulphuric acid (5 cm^3) to concentrated nitric acid (5 cm^3), and cool the mixture to room temperature. While shaking the flask, slowly add the unknown (0·5 g). If the unknown is a liquid, and dissolves quickly, allow the mixture to stand at room temperature for 15 minutes then withdraw a small portion into water: if a solid is produced, pour the bulk into water (50 cm^3). If no solid is produced, heat the mixture for 30 minutes at 60° and again withdraw a sample into water. If there is still no solid, heat the mixture below boiling point for 5 minutes, cool, and pour slowly into water (50 cm^3). Filter off any solid residue and recrystallize from ethanol: compare the m.p., etc., with the original unknown and with the nitro derivatives listed in the tables.

If nitration has not occurred try method (ii).

(ii) Dissolve the unknown (0·5 g) in glacial acetic acid (10 cm^3), and cool the solution in ice; slowly add fuming nitric acid (5 cm^3).

HANDLE FUMING NITRIC ACID WITH CARE.

Warm the mixture in a boiling water bath for 15 minutes, and withdraw a sample for test as in (i). If no reaction has occurred, bring to boiling point for 5 minutes, cool, and again test. If further reaction is needed, bring to boiling and maintain the mixture near boiling point for 15 minutes before cooling and slowly pouring into water. Work up as in (i).

If nitration has not occurred try method (iii).

(iii) Slowly add concentrated sulphuric acid (5 cm^3) to fuming nitric acid

(5 cm^3), and cool the solution in ice. Slowly add the unknown, with constant swirling of the mixture, and gradually raise the temperature of the reaction to 100° in a boiling water bath; withdraw samples for test as in (i). Maintain the reaction at 100° until nitration has occurred, or for 30 minutes, then cool and pour slowly into water (50 cm^3). Work up as in (i).

Bromo Derivatives (Aromatic Ethers)

Dissolve or suspend the unknown (0·5 g) in chloroform (5–10 cm^3), and add a solution of bromine in chloroform slowly, as fast as the bromine colour is consumed. Warm in a water bath at 60°, and add more bromine in chloroform until the bromine colour persists. Distil off the chloroform, wash the product with dilute aqueous sodium bicarbonate, and recrystallize from ethanol.

Side-Chain Oxidation (Aromatic Ethers and Hydrocarbons)

Ethers and hydrocarbons with an alkyl side-chain can be oxidized to the corresponding aryl carboxylic acid by strong alkaline permanganate. The m.p. of these acids are given in table 4.18.

In a round bottom flask place 50 cm^3 of 10 per cent sodium carbonate solution, the unknown (1·5 g), and potassium permanganate (4 g). The mixture must now be refluxed: some hydrocarbons are fully oxidized within 30 minutes, some require many hours for complete oxidation. In most instances, substantial amounts will be oxidized after 1 hour, and this is recommended as reasonably general.

Cool the mixture, and slowly add solid sodium bisulphite (5 g), then boil the mixture for a further 5 minutes, and filter off any manganese dioxide. Again cool the filtrate, and acidify strongly with concentrated hydrochloric acid (CO_2!). Cool the mixture in ice/salt, and filter off the free acid: if no acid precipitates at 0°, extract with ether. Recrystallize the acid from water or aqueous ethanol. Dialkylbenzenes usually give the dicarboxylic acid.

Alkanes, Alkenes, and Alkynes

No generally useful derivatives are available for these classes; the most reliable methods of identification involve gas chromatographic and mass spectrometric methods (chapter 7). The n.m.r. spectrum is also important, especially for simple members.

4.2f Carboxylic Acids

Equivalent Weight
(i) Dissolve the acid (*ca.* 0·2 g, accurately weighed) in ethanol (25 cm^3), and titrate with 0·1M sodium hydroxide using phenolphthalein as indicator. Simultaneously carry out a duplicate and a blank. Alternatively, dissolve the acid in standard sodium hydroxide and back-titrate the excess with standard acid.

(ii) If a sharp end-point cannot be obtained in (i), dissolve the acid in dimethylformamide (25 cm^3) or benzene + methanol (20 + 5 cm^3) and titrate with 0·1M sodium methoxide in benzene–methanol, using thymol blue or azo violet as indicator. (Prepare the sodium methoxide by gradually adding clean sodium metal (2·5 g) to methanol (100 cm^3) in a 500 cm^3 flask. CAUTION: HYDROGEN EVOLVED. When the sodium has dissolved, make up with benzene–methanol (4:1); more methanol may be added to keep the solution clear. Standardize before use with benzoic acid.) Use fresh solutions, or protect them from atmospheric CO_2 with soda-lime tubes. Run a simultaneous duplicate and a blank titration.

S-Benzylthiuronium Salts
Dissolve the acid (0·5 g) in the minimum quantity of dilute sodium hydroxide solution: use a drop of phenolphthalein indicator to check for excess. Now add dilute hydrochloric acid dropwise until the phenolphthalein colour is just discharged; the resulting solution is a substantially neutral solution of the sodium salt of the acid. Warm the solution almost to boiling, and add solid S-benzylthiuronium chloride (1 g), stirring continuously; on slow cooling, the derivative will crystallize out. Filter off, and wash on the filter with a little ice-water; dry on filter paper or in a vacuum desiccator.

Do not attempt to recrystallize. Do not dry in the oven. Both of these operations are likely to decompose the derivative to benzyl mercaptan, a compound with an unforgettable smell.

$$PhCH_2S-C\overset{NH_2{}^+Cl^-}{\underset{NH_2}{\diagdown}} + RCOO^- \longrightarrow PhCH_2S-C\overset{NH_2{}^+(RCOO^-)}{\underset{NH_2}{\diagdown}} \xrightarrow[OH^-]{heat} PhCH_2SH$$

Anilides
(i) Prepare the acid chloride of the acid by refluxing the unknown (0·5–1·0 g) with thionyl chloride (5 cm^3) and a few drops of pyridine for 30 minutes: carry out this reaction in the fume cupboard. Remove excess thionyl chloride by distillation (b.p. 79°), then add benzene (5 cm^3) and distil this off to remove any remaining thionyl chloride.

(ii) To the crude acid chloride from (i) add aniline (1 cm^3): the anilide is usually formed immediately, but if in doubt warm on the water bath for 10 minutes. Cool in ice if the material is not crystalline, then filter off and recrystallize from ethanol or aqueous ethanol.

$$RCO_2H \longrightarrow RCOCl \longrightarrow RCONHPh$$

p-Toluidides
Proceed as for anilides, substituting p-toluidine for aniline.

$$RCO_2H \longrightarrow RCOCl \longrightarrow RCONHC_6H_4CH_3$$

Amides
Proceed as for anilides, but treat the crude acid chloride dropwise with 0·88 ammonia. Recrystallize preferably from absolute ethanol, in which inorganic byproducts are insoluble.

$$RCO_2H \longrightarrow RCOCl \longrightarrow RCONH_2$$

p-Bromophenacyl Esters
Proceed as above (preparation of S-benzylthiuronium salts) to obtain a neutral solution of the salt of the acid. Add a solution of p-bromophenacyl bromide (0·5 g) in ethanol (5 cm³) and reflux for 1 hour for monobasic acids (2 or 3 hours for di- or tri-basic acids respectively). Cool until the derivative crystallizes, filter off and recrystallize from ethanol.

The reagent is extremely irritating to skin and eyes: handle with care. Carry out the reflux in a fume cupboard.

p-Nitrobenzyl Esters
These are prepared as for p-bromophenacyl esters, but substituting p-nitrobenzyl bromide for p-bromophenacyl bromide. (In general the same precautions in handling the reagent should be practised.)

4.2g Carboxylic Anhydrides

Equivalent Weight
Treat the anhydride (0·2 g) with excess standard alkali and back titrate with standard acid (as in 4.2f).

Anilides, p-Toluidides, and Amides
These can be prepared directly from the anhydride by treating it with aniline, p-toluidine, or ammonia: adapt the methods given in 4.2f. In the

case of lower aliphatic anhydrides, this method is not a very efficient route to amides because of purification difficulties.

Hydrolysis
The anhydride can be hydrolysed by water (or alkali) to the acid (or its salt), and identified by any of the methods given for acids in 4.2f. Particularly easy are the S-benzylthiuronium salts, and p-bromophenacyl esters.

4.2h   Phenols and Enols

Enols should be treated as ketones or esters as the case may be.

Equivalent Weight
The method given for acids, using non-aqueous reagents, is suitable for phenols: see 4.2f (page 68).

Aryloxyacetic Acids
Reflux together the phenol (1 g) with chloroacetic acid (1·2 g) and 2M sodium hydroxide solution (10 cm^3) for 30 minutes. Cool, and acidify carefully with concentrated hydrochloric acid: cool in ice/salt until the derivative crystallizes out. If no crystals are formed, extract the solution with a little ether, wash this with water and extract the aryloxyacetic acid from the ether with dilute sodium carbonate solution (10 cm^3). Acidify the carbonate carefully with concentrated hydrochloric acid (CO$_2$!), filter off the derivative, and recrystallize from water. The equivalent weight of the derivative can be determined (as outlined in 4.2f) as well as m.p.
    The method does not succeed with nitrophenols.

$$PhO^- + CH_2CO_2^- \longrightarrow PhO.CH_2CO_2^- \xrightarrow{H^+} PhO.CH_2CO_2H$$

Benzoates (the Schotten–Baumann Benzoylation Reaction)
In a 50 cm^3 flask place 2M sodium hydroxide solution (20 cm^3) and the unknown phenol (1 g): add benzoyl chloride (2 cm^3), cork the flask firmly, and shake it for 10 minutes. Release the cork occasionally: carry out this reaction in the fume cupboard. After 10 minutes, the smell of benzoyl chloride should be substantially gone.
    If the derivative is not solid, cool in ice with continuous shaking then filter off and recrystallize from ethanol or aqueous ethanol. Simple phenols give fairly low melting benzoates, and they are not recommended for this group.

$$Ph-\underset{\underset{Cl}{}}{\overset{O}{\underset{\|}{C}}}-OPh \longrightarrow Ph-\overset{O}{\overset{\|}{C}}-OPh + Cl^-$$

p-Nitrobenzoates
Proceed as for alcohols in 4.2d (page 63).

3,5-Dinitrobenzoates
Proceed as for alcohols in 4.2d (page 62).

Toluene-p-sulphonates, Tosylates
Reflux a mixture of phenol (1 g), toluene-p-sulphonyl chloride (1·5 g), and pyridine (3 cm^3) for 15 minutes. Cool, pour onto a mixture of ice and concentrated hydrochloric acid, and continue cooling in ice until the derivative crystallises: filter off and recrystallise from ethanol.

1-Naphthylurethanes
Proceed as for alcohols in 4.2d (page 63), but first add a few drops of pyridine to the reaction mixture.

Bromo Derivatives
Proceed as for aromatic ethers in 4.2e (page 67). The reaction can also be carried out in aqueous solution, and using bromine–water; the bromo derivative usually separates out during the course of the reaction.

4.3a–4.3g   All Amines

Equivalent Weight
Most amines are too weakly basic to be titrated in aqueous media: they can be titrated in glacial acetic acid with perchloric acid (also in glacial acetic acid) using methyl violet or crystal violet as indicator. Perchloric acid is obtained as an aqueous solution, and acetic anhydride must be added to remove the water. In a litre flask or bottle mix together perchloric acid solution (72 per cent, 8·5 cm^3) and glacial acetic acid (ca. 500 cm^3): add acetic anhydride (20 cm^3) and shake the mixture for several minutes. Make up to approximately 1 litre with glacial acetic acid, and allow the solution to stand overnight. The solution should be protected from atmospheric moisture by careful stoppering or by a silica gel guard tube. Standardize the perchloric acid before use with potassium hydrogen phthalate (ca. 0·5 g) dissolved in 50 cm^3 of glacial acetic acid; use a few drops of indicator (0·5 per cent in glacial acetic acid), and titrate till the colour changes from violet to blue-green (but not yellow). Titrate the amine (ca. 0·3 g) also in 50 cm^3 of glacial acetic acid. Simultaneously, run a duplicate and a blank estimation.

Details for the preparation of many derivatives for these classes have been given above: any changes are noted here: for tertiary amines, picrates are the most satisfactory.

Benzamides

Use the Schotten–Baumann benzoylation reaction as for the preparation of phenolic benzoates, 4.2h, page 70. For amino acids (e.g., anthranilic acid) do not use an excess of benzoyl chloride, since the *derivative is an acid* and must be liberated from the alkaline solution by acidification; excess benzoyl chloride goes to benzoic acid, which is difficult to separate from the derivative.

$$\text{PhCOCl} + \text{RNH}_2 \longrightarrow \text{PhCONHR}$$

Toluene-*p*-sulphonamides:

Use the method given for the tosylates of phenols, 4.2h, page 71. Tosyl derivatives can also be prepared by the Schotten–Baumann reaction (substituting toluene-*p*-sulphonyl chloride for benzoyl chloride): if this method is used, *derivatives for primary amines form soluble sodium salts* and must be liberated from alkaline solution by acidification with concentrated hydrochloric acid.

Primary: $\text{CH}_3.\text{C}_6\text{H}_4.\text{SO}_2\text{Cl} + \text{RNH}_2 \longrightarrow \text{CH}_3.\text{C}_6\text{H}_4.\text{SO}_2\text{NHR}$

$\text{CH}_3.\text{C}_6\text{H}_4.\text{SO}_2\text{N}^-\text{R} \xleftarrow{\quad \text{OH}^- \quad}$

Secondary: $\text{CH}_3.\text{C}_6\text{H}_4.\text{SO}_2\text{Cl} + \text{R}_2\text{NH} \longrightarrow \text{CH}_3.\text{C}_6\text{H}_4.\text{SO}_2\text{NR}_2$

Picrates

Use the method given for ethers, etc., 4.2e (page 65).

Acetamides

(i) Most amines are acetylated by the following method: shake the amine in a test-tube with water (3 cm^3) and acetic anhydride (1 cm^3) for 5 minutes. Warm the tube in a water bath until the excess anhydride is consumed, then, still shaking the tube, cool in ice. Filter off and recrystallize from water or aqueous ethanol.

(ii) All primary and secondary amines, their salts, and all OH compounds, are acetylated by the following: boil the unknown (0·5 g) with acetic anhydride (1 cm^3) and anhydrous sodium acetate (0·5 g) for 5 minutes. If the unknown does not dissolve, add 3 cm^3 of pyridine and boil for 5 minutes. Cool, and pour into a slight excess of cold dilute sodium hydroxide: cool in ice to recrystallize the derivative, then filter off and recrystallize from water or aqueous ethanol.

2,4-Dinitrophenyl Derivatives

Dissolve the amine (0·5 g) and 2,4-dinitrochlorobenzene (more correctly named chloro-2,4-dinitrobenzene) (0·5 g) in the minimum of boiling ethanol, and add anhydrous sodium acetate (0·5 g): reflux for 30 minutes. Cool in ice, and add cold water until the derivative is fully precipitated: filter off and recrystallize from ethanol. 2,4-Dinitrochlorobenzene must be handled with care, as it blisters the skin. Wash off any accidental spillage immediately with a little ethanol followed by soap and water.

Methiodides (Tertiary Amines)

In a dry test-tube, treat the amine (0·5 g) with methyl iodide (1 cm^3): stir continuously while warming on a water bath for 5 minutes. Cool in ice, while scratching the walls of the tube with a glass rod: crystallization is often difficult, and the mixture may have to be kept at low temperature for some time. Filter off the methiodide and recrystallize from absolute ethanol (not 95 per cent), ethyl acetate, or ethanol–ether.

$$R_3N + CH_3I \longrightarrow [R_3\overset{+}{N}CH_3] I^-$$

p-Nitroso Derivatives (Tertiary Aromatic Amines)

Dissolve the tertiary amine (0·5 g) in dilute hydrochloric acid (10 cm^3) and cool to below 5°. Slowly add sodium nitrite, until a slight excess is present: test with starch–iodide paper. Now liberate the dark green p-nitroso base from its hydrochloride by adding dilute sodium hydroxide until the solution is distinctly alkaline, and extract the derivative with ether. Distil off the ether, and recrystallize from light petroleum (b.p. 60–80°) or benzene.

p-Nitroso tertiary amines usually crystallize in glistening dark green plates.

red   dark green

4.3h Hydrazines and Semicarbazides

Equivalent Weight
These can be titrated in glacial acetic acid against acetous perchloric acid as for amines: see 4.3a, etc.

Acetophenone (PhCOMe) Derivatives
For hydrazines, use the method given for the preparation of aldehyde *p*-nitrophenylhydrazones (page 58).

For the hydrochlorides of phenylhydrazine or semicarbazide use the aldehyde semicarbazone method (page 58).

Any ketone or aldehyde may be chosen to react with these hydrazines or semicarbazides, the melting points in most cases being given in the aldehyde and ketone tables.

4.3i Imines, Schiff's Bases, and Aldehyde-Ammonias

Prepare the 2,4-dinitrophenylhydrazone of the constituent carbonyl compound by the method given for aldehydes on page 57. In case of ambiguity the *p*-nitrophenylhydrazone might also be prepared.

4.3j Primary Amides

Hydrolysis
The surest means of identifying an amide is to hydrolyse it to the corresponding acid, which is then identified as in 4.2f.

Reflux the amide (1 g) with 30 per cent sodium hydroxide solution (5–10 cm^3) until no more ammonia is given off (usually 30 minutes): cool, and very carefully acidify with hydrochloric acid (begin with dilute, then use concentrated). Cool in ice, filter off any precipitated acid, or extract soluble acids with ether. Aromatic acids usually crystallize out in the acid solution. Identify the acid normally, repeating the Lassaigne's test.

Xanthylamides

Dissolve xanthydrol (0·5 g) in glacial acetic acid (7 cm^3), decanting the clear solution from any insoluble material. Dissolve the amide (0·5 g) in this solution, adding a little ethanol if necessary, then heat the mixture on a water bath for no more than 30 minutes. Cool in ice if necessary, and dilute with water if necessary: filter off the xanthylamide and recrystallize from aqueous acetone, aqueous dioxan, or aqueous acetic acid. Dry on filter paper or in a vacuum desiccator.

4.3k N-Substituted Amides

These are difficult compounds to identify, as they must be hydrolysed to the constituent acid and amine, each of which should be identified as a separate unknown compound: in many cases, identification of *either* the acid *or* the amine may be sufficient.

Consult the various lists of substituted amides (see above table 4.34) and reach a tentative conclusion about the identity of the compound. If the amide is formed from an aromatic amine and an aliphatic acid, concentrate on identifying the amine (it is usually easier to isolate an aromatic compound): if formed from an aliphatic amine and aromatic acid, concentrate on identifying the acid.

Reflux the unknown (2 g) with *ca.* 67 per cent sulphuric acid (5 cm^3 of concentrated acid added slowly to 5 cm^3 of water): hydrolysis will take 30 minutes to 1 hour. Cool the solution; it contains the free organic acid and the sulphate of the amine.

If the acid is aromatic, it may crystallize out on cooling in ice; if so, dilute the solution a little and filter off. Otherwise extract the acid with a little ether, and distil off the ether to try and isolate the acid. Some acids are water soluble and insoluble in ether; these will not be isolated by this procedure; see below.

Having removed the organic acid, make the residual solution strongly alkaline with 30 per cent sodium hydroxide solution; CAUTION! Either try to steam distil the amine (it is sufficient to distil over a part of the aqueous solution directly), or extract it with ether. If steam distillation has been used, extract the amine from the steam distillate with ether. In either case, distil off the ether and identify the amine: it may not be easy to purify the amine, but it will be sufficiently pure for the preparation of a derivative (e.g., benzoyl): see 4.3a–4.3g.

Note: Substituted amides of the following acids do not permit the easy isolation of the free acid: carbonic acid (ureas, urethanes), oxalic acid, succinic acid, acetoacetic acid (decarboxylates rapidly).

It is usually sufficient to identify the amine in these cases.

Note also that amides of acetoacetic acid are enolic, and chemical and spectroscopic evidence will be adduced for this: see class 2h.

$$R-\underset{\underset{}{\overset{\overset{O}{\parallel}}{}}}{C}-NHR' \xrightarrow{H^{+}} R-\underset{\underset{\overset{\cdot\cdot}{\underset{H}{O}}\diagdown H}{\overset{\overset{\diagup H}{O}}{}}}{\overset{+}{C}}-NHR' \longrightarrow \longrightarrow R-C\underset{OH}{\overset{O}{\diagup\kern-0.6em\diagdown}} + R'\overset{+}{N}H_3$$

4.3l   Ammonium Salts of Carboxylic Acids

These are identified by simple modifications of the methods used for the parent acid: see 4.2f. Ammonium salts can be used directly to prepare S-benzylthiuronium salts, etc.

4.3m   Aminophenols

In general these show simple phenolic and amine properties, although they are often quite strong reducing agents: acetylation, benzoylation or toluene-p-sulphonylation are the safest derivatives to try, remembering that these processes can affect both the phenol and the amine group. Details are given under phenols 4.2h. They cannot be titrated to visual end-points because they themselves produce coloured material during the titrations.

4.3n   Amino Acids

Amino acids having an aliphatic amine group exist as zwitterions: many are optically active, but the specific rotation often changes according to pH, and this can only be used for identification purposes where comprehensive information on such changes is available.

Equivalent Weight
They can be titrated as bases using the acetous perchloric acid method outlined for amines (4.3a, etc.). Do not use heat to dissolve the amino acids in the acetic acid: low values may result, possibly because of acetylation of the amino group.

Benzoylation, 3,5-Dinitrobenzoylation, and
Toluene-*p*-sulphonylation
These processes can be carried out using the selection of methods given for
phenols (see 4.2h). EXCESS OF THE REAGENT MUST BE AVOIDED. The derivative produced is itself acidic, and must be liberated from alkaline solution
by acidification: excess benzoyl chloride (etc.) will produce benzoic acid
(etc.) in this process, and this will make purification difficult.

Paper Chromatography
Paper chromatography (or T.L.C.) of unknown amino acids against known
standards is the best method of identification. Suitable paper is Whatman
No. 1: a suitable mobile phase is prepared by mixing propan-2-ol and
dilute ammonia solution (2:1). Develop the chromatogram for 1 hour,
dry, and spray with ninhydrin reagent (1 per cent in ethanol): full colours
will be developed when the chromatogram is placed in an oven at 100° for
5 minutes.

4.3o   Nitriles

As with amides, the only general method for the identification of nitriles
involves hydrolysis to the acid.

Carry out the hydrolysis with 30 per cent aqueous sodium hydroxide
or 67 per cent sulphuric acid: the former usually works best for aliphatic
members, while the acid hydrolysis is preferred for aromatic nitriles.
Isolation of the acid follows normal procedures: read the sections on
alkaline hydrolysis of primary amides (4.3j) and on acid hydrolysis of
substituted amides (4.3k).

$$R-C\equiv N \longrightarrow RCO_2H + NH_3$$

4.3p   Azo Compounds

Symmetrical azo compounds (e.g., 2,2′-dimethylazobenzene) are reduced
by metal and acid to the corresponding amine (e.g., *o*-toluidine) *Unsymmetrical* azo compounds give a mixture of amines: the simplest general
way to identify these is to prepare an authentic sample of the probable
compound and compare the physical properties with the unknown
(infrared spectrum, behaviour on T.L.C., etc.).

Reduction
Reflux the unknown (2 g) for 30 minutes with tin (2 g) and hydrochloric
acid (20 per cent, 50 cm³). Cool, and make alkaline with 30 per cent sodium
hydroxide solution. Distil over 25 cm³ of aqueous amine, and prepare a

benzoyl (or tosyl) derivative of the amine directly on the aqueous steam distillate. Thereafter refer to the tables of amines. See also the notes on amines.

$$R—N{\equiv}N—R \xrightarrow[\text{metal}]{M} \underset{\text{radical anion}}{R—\overset{.}{N}—\overset{-}{N}—R} (+ M^+) \xrightarrow{H^+} R—\overset{.}{N}—\overset{\overset{\displaystyle H}{|}}{N}—R$$

$$R—\overset{\overset{\displaystyle H}{|}}{N}—\overset{.}{N}—R \xrightarrow{M} R—\overset{\overset{\displaystyle H}{|}}{N}—\overset{-}{N}—R (+ M^+) \xrightarrow{H^+} R—\overset{\overset{\displaystyle H}{|}}{N}—\overset{\overset{\displaystyle H}{|}}{N}—R \xrightarrow{\text{etc.}}$$

$$RNH_2 + H_2NR$$

4.3q   Nitroso Compounds

C-nitroso compounds are commonly the nitroso derivatives of amines and phenols: the latter exist as the isomeric quinone monoximes, while for amines the corresponding equilibrium (where appropriate) lies well over to the true nitroso form. Identification is not easily put on a general basis, but *reduction* of the amine nitroso compounds as for azo compounds is useful: nitroso derivatives of phenols can be *hydrolysed* in acid to the corresponding quinone, which can then be isolated and examined separately.

4.2e   Aromatic Nitro Hydrocarbons and Ethers

Reduction
Reduction of the nitro group by metal and acid gives the corresponding primary amine. Use the details given for azo compounds on page 77.

Reductive Acetylation
Isolation of the free amine formed by reduction is often troublesome: reduction in the presence of acetic anhydride gives the acetyl derivative of the amine, which is more easily isolated. Use the details given in 4.2a for the preparation of hydroquinone acetates from quinones.

$$RNO_2 \xrightarrow[H^+]{M} RNH_2 \xrightarrow{(CH_3CO)_2O} RNHCOCH_3$$

4.3r   Imides, including Cyclic Urea Derivatives

Equivalent Weight
Imides and barbituric acids can be titrated in dimethylformamide solution against methanolic sodium methoxide, using thymol blue as indicator: the method is the same as for acids (4.2f).

Hydrolysis

Imides should be hydrolysed to the corresponding acid, as described for amides (page 74): slightly longer hydrolysis times should be used.

Barbituric acid on hydrolysis will give the same products as a mixture of malonic acid and urea, i.e., acetic acid, ammonia, and CO_2: with alkaline hydrolysis the ammonia should be detected coming off, while the CO_2 will give carbonate which will liberate CO_2 on acidification. Disubstituted barbituric acids give substituted malonic acids:

$$R_2C\begin{array}{c}CO-NH\\CO-NH\end{array}CO \longrightarrow 2NH_3 + CO_2 + R_2C\begin{array}{c}CO_2H\\CO_2H\end{array} \longrightarrow R_2CHCO_2H + CO_2$$

4.4a Acyl Halides

Hydrolysis

All acid chlorides are hydrolysed by water (alkali catalyses the reaction) to the corresponding acid: acetyl chloride reacts vigorously while benzoyl chloride reacts fairly slowly, and these can be taken as models for the others.

Allow the acid chloride (1 g) to react with water (5 cm³) either in the cold (lower aliphatics) or boiling (aromatics) for 10 minutes. If the acid crystallizes out on cooling, filter off and identify as a member of class 2f. If it does not crystallize out, prepare derivatives from the aqueous solution (see 4.2f), e.g., the S-benzylthiuronium salt. Chloroformate esters are best identified by conversion to the amide, i.e., the carbamate ester, or 'urethane'. (See below.)

Amides

Convert the acid chloride to the acid amide by treatment with 0·88 ammonia solution (see 4.2f). Chloroformates give urethanes, so that ethyl chloroformate gives ethyl carbamate (urethane m.p. 49°), and methyl chloroformate gives methyl carbamate (m.p. 52°): these must be extracted from the aqueous reaction mixture with ether, and recovered by distillation of the ether.

$$RCOCl \xrightarrow{H_2O} RCO_2H + HCl$$
$$RCOCl \xrightarrow{2NH_3} RCONH_2 + NH_4Cl$$
$$Cl.COOCH_3 \xrightarrow{2NH_3} H_2N.COOCH_3$$
$$\text{methyl carbamate}$$
$$\text{(a urethane)}$$

4.4b   Alkyl Halides and Aromatic Side-Chain Halides

S-Alkylthiuronium Picrates
Reflux the unknown (1 g) with thiourea (1 g) in ethanol (10 cm³) for 20
minutes. Add picric acid (1 g) and again reflux briefly. Cool, and filter off
the picrate. Try to recrystallize a few mg in a test-tube to ensure that the
picrate can be recrystallized (from ethanol): if this is satisfactory, repeat
with the bulk of the derivative.

Picrate of the Alkyl 2-Naphthyl Ether
Reflux for 15 minutes a mixture of the unknown (1 g), 2-naphthol (2 g),
potassium hydroxide (1 g) and ethanol (10 cm³). Cool the mixture and
dilute with 2M potassium or sodium hydroxide (20 cm³). If the 2-naphthyl
ether crystallizes out, filter it off: otherwise extract it with ether (5 cm³).
Prepare a saturated solution of picric acid (0·5 g) in ethanol at the boiling
point, and to this add the 2-naphthyl ether (either dissolved in the minimum of ethanol or as the ether extract obtained above). Bring the united
solutions near boiling point for a minute, then cool, filter off the picrate,
and wash with a very little ethanol. Stop the washing if the picrate
changes colour, indicating dissociation to picric acid and the free 2-
naphthyl ether.

4.4c.   Aryl Halides

Nitration
This is carried out as described for ethers and hydrocarbons in 4.2e:
method (iii) is most often successful.
    EXERCISE THE SAME PRECAUTIONS GIVEN IN 4.2e.

The Grignard Reaction
Most of the monohalogeno compounds listed in table 4.45 can be converted
smoothly to the Grignard reagent, which reacts with CO_2 to give a solid
carboxylic acid. Very brief details are given here.
    All apparatus and reagents must be thoroughly dry.
    In a 50 cm³ round bottom flask place magnesium turnings (suitable for
Grignard use, 0·3 g) and anhydrous ether (10 cm³): add a crystal of iodine,
the size of a rice grain. Dissolve the aryl halide (2 g) in anhydrous ether
(5 cm³) and add half of this solution to the flask. Reflux the reaction mixture
(NO FLAMES NEARBY) until the solution becomes cloudy and the reaction
becomes self supporting. Add the remainder of the aryl halide solution,
and keep the reaction going, if necessary with further heating on the water
bath, until almost all the magnesium has reacted. Slowly add 10 g of
crushed solid CO_2, or pass CO_2 gas through the reaction mixture for
15 minutes. Finally acidify the mixture strongly with hydrochloric acid
(5M, 15 cm³), and steam distil off the ether: this can be done by *preheating*
an oil bath to 150°, and using this to heat the reaction flask while distilling
off the ether. Cool the reaction mixture, and filter off the acid: recrystallize from water or ethanol, and identify as in 4.2f.

4.5a   Mercaptans, Sulphides, Disulphides, and
Thioacids

Derivatives are only given for the mercaptans: others can usually be
identified from their physical properties.

2,4-Dinitrophenylthioethers
Reflux the mercaptan (0·5 g) in ethanol (20 cm³) with a little sodium
hydroxide solution (2M, 2 cm³) and 2,4-dinitrochlorobenzene (1 g). Handle
the reagent with care: it blisters the skin: see also page 73. Continue the
reflux for 15 minutes, then cool in ice. Filter off the precipitated thioether
(sulphide) and recrystallize from ethanol.

Hydrolysis of Thioacids

Reflux the thioacid (1 g) with dilute sodium hydroxide solution ($10\,cm^3$) for 15 minutes. Cool, add 20 volume hydrogen peroxide ($2\,cm^3$) and reflux for a further 10 minutes. The solution now contains the anion of the normal carboxylic acid: identify the acid as in 4.2f. Preparation of the S-benzyl-thiuronium salt on the neutralized hydrolysis solution may be the most convenient.

$$\left(R{-}C\!\!\underset{SH}{\overset{O}{\diagup}} \rightleftharpoons R{-}C\!\!\underset{S}{\overset{OH}{\diagup}} \right) \xrightarrow{\;H_2O\;} R{-}C\!\!\underset{OH}{\overset{O}{\diagup}} + H_2S$$

4.5b   Sulphonic Acids

The free acids are not frequently met, and are seldom pure enough to justify measuring their equivalent weights. The m.p. or decomposition points of sulphonic acids and their salts are of no value in identification.

S-Benzylthiuronium Salts

The method is exactly the same as for carboxylic acids (4.2f). In table 4.47 the acids are arranged according to the m.p. of these salts.

Sulphonyl Chlorides

Heat the acid or its salt (1 g) with phosphorus pentachloride (2 g) for 30 minutes at $150°$: use an oil bath or glycerol bath. Cool, and extract the acid chloride thoroughly with several portions of benzene (total $20\,cm^3$). Distil off the benzene, latterly under reduced pressure and recrystallize the residue from benzene or benzene-light petroleum (b.p. 60–$80°$)

$$RSO_2OH + PCl_5 \longrightarrow RSO_2Cl + POCl_3 + HCl$$

Sulphonamides

Treat the concentrated benzene solution of the acid chloride (obtained above after most of the benzene has been distilled off) with 0.88 ammonia solution ($10\,cm^3$). Agitate the mixture continuously for 5 minutes until the sulphonamide crystallizes out: if it does not, remove the benzene by distilling it azeotropically from the reaction mixture, and cool the aqueous residue in ice. Filter off the sulphonamide and recrystallize from ethanol or aqueous ethanol. The equivalent weight of the sulphonamide can then be measured: see 4.7c. Use the non-aqueous method, with dimethylformamide as solvent.

$$RSO_2Cl + 2NH_3 \longrightarrow RSO_2NH_2 + NH_4Cl$$

Sulphonanilides

Reflux the concentrated benzene solution of sulphonyl chloride obtained above with aniline ($1 \, cm^3$) for 1 hour. Distil off the benzene until the anilide begins to crystallize: cool in ice, filter off, and recrystallize from ethanol or aqueous ethanol.

$$RSO_2Cl + 2PhNH_2 \longrightarrow RSO_2NHPh + PhNH_3^+Cl^-$$

4.5c   Sulphonate Esters

These can be hydrolysed as for carboxylic esters (see 4.2b), but make no attempt to isolate the free sulphonic acid: isolate the alcohol or phenol as usual, and prepare the S-benzylthiuronium salt of the acid directly on the neutralized solution of the sodium salt.

4.5d   Sulphate Esters

Dialkyl sulphates, especially dimethyl sulphate, are toxic. You would not normally be given such a compound as an unknown without adequate warning of the toxicity.

Monoalkyl sulphates and their salts are fairly common.

Hydrolysis

Follow the details given for carboxylic esters (4.2b): isolate the alcohol or phenol as usual and identify separately, and confirm the presence of SO_4^{2-} in the hydrolysate.

4.6a   Halogen Substituted Nitro Hydrocarbons and Ethers

There is no general set of derivatives for these compounds, since the reactivity of the halogen atoms is so variable. In the case of mononitro compounds, an attempt should be made to reduce them to the halogeno amine (see 4.2e and 4.3p). Compounds with the halogen atom in the side-chain can be oxidized easily to the corresponding nitro carboxylic acids using the method given for hydrocarbons (4.2e), but oxidation proceeds much faster (30 minutes).

Compounds with 2 or 3 nitro groups on the nucleus can be hydrolysed with aqueous sodium carbonate (cold or hot) to the nitrophenols: reaction progress can be followed by the development of the intense yellow colour of the nitrophenate ions. Isolate the phenols by acidification and extraction with ether or benzene.

4.7b Thioureas and Thioamides

Primary thioamides on hydrolysis give ammonia and the thioacid, which further hydrolyses to the normal carboxylic acid: this should be isolated by the usual methods (see the hydrolysis of carboxylic amides, 4.3j). (Note that thioacids are usually an equilibrium mixture of $R.CS.OH$ and $R.CO.SH$.)

Thioanilides on hydrolysis give primary aromatic amines: use the method given for the hydrolysis of carboxylic acid anilides (N-substituted amides) in 4.3k.

N-substituted thioureas on hydrolysis also give amines, which can be isolated and identified as in 4.3k.

Thiourea itself gives ammonia and CO_2.

4.7c Sulphonamides and N-Substituted Sulphonamides

Equivalent Weight
Primary and secondary sulphonamides are weakly acidic, and can be titrated as such in non-aqueous media: use the method given for carboxylic acids (4.2f) using dimethylformamide as solvent, sodium methoxide as base, and thymol blue as indicator.

Hydrolysis
Sulphonamides of all types are difficult to hydrolyse: a few are successfully treated by the method given for N-substituted carboxylic amides (4.3k). All are hydrolysed by the following. In a large test-tube place water (0.5 cm^3) then phosphoric acid (85 per cent, 2 cm^3) then concentrated sulphuric acid (2 cm^3). Add the sulphonamide (2 g) and heat over a small flame, with constant shaking, until the temperature reaches 150°. CAUTION: maintain the mixture at this temperature until the sulphonamide has reacted, and the mixture is homogeneous. Cool thoroughly, then pour on to ice (10 g).

Isolate and identify the amine as outlined in 4.3k. Identify the sulphonic acid by direct conversion to the S-benzylthiuronium salt (see 4.3k).

4.7d Aminosulphonic Acids

The presence of the amine group precludes the preparation of the acid chloride, and hence the direct preparation of sulphonamides, etc.

The equivalent weight should be determined as for carboxylic acids (4.2f), and the S-benzylthiuronium salt prepared normally (see 4.2f).

Tables of physical constants

Nomenclature

Throughout the tables, systematic nomenclature has been used together with the commoner trivial names; it is always difficult to decide when to stop using trivial names and start using systematic names, and in these tables it is most frequently done around the C_3 member of a series. Thus in table 4.1 the name acetaldehyde is used, but propanal is given for the next member.

The prefix n- has frequently been dropped from the names of unbranched alkyl groups (propyl, butyl, etc.), but it is equally correct to use the prefix.

The British custom is to insert numbers immediately before the function which they qualify (butane-2,3-dione); this differs from the American usage (2,3-butanedione), but does not in any way produce ambiguity. There is a current tendency towards the latter form, and it is to be hoped that a truly universal agreement will be reached soon.

The use of hyphens is particularly troublesome. Since the *Chemical Abstracts* recommendations are familiar to all chemists (and since this means the use of the least possible number of hyphens) this is the practice adopted in this book.

Further information and reasoned opinion can be found in R. S. Cahn, *An Introduction to Chemical Nomenclature*, Butterworths, London, 4th Ed., 1974.

Additional Reference Sources

1. *Handbook of Chemistry and Physics* (The Chemical Rubber Co., Ohio, 47th Ed., 1967) lists compounds according to important functional classes, and contains a formula index. Very comprehensive lists of derivatives are given. (In English)

2. *Tables for Identification of Organic Compounds* (2nd Ed., 1964), is a supplement to the larger *Handbook of Chemistry and Physics*. It does not contain a formula index. (In English)

3. *Dictionary of Organic Compounds* (Eyre and Spottiswood, 5 volumes plus annual supplements), is an alphabetically arranged list of important organic compounds, giving data on physical properties and derivatives. Usually referred to as 'Heilbron', after the late chairman of the editorial board, Sir Ian Heilbron. (In English)

4. *Beilstein's Handbuch der Organischen Chemie* sets out to catalogue all organic compounds: the main work (*Hauptwerk*) covers the years to 1910; the first supplement (*Erstes Ergänzungswerk*) to 1919; the second supplement (*Zweites Ergänzungswerk*) to 1929; and the third supplement (*Drittes Ergänzungswerk*), not yet complete, to 1949.

A name index and a formula index appear at the end of the *Zweites Ergänzungswerk*, indexing the first two supplements and the *Hauptwerk*; in addition, each volume contains its own name index. The work is divided into 27 volumes (Band I–Band XXVII), each volume dealing with specified classes of compound, e.g., acyclic monocarboxylic acids, isocyclic monoamines. Physical constants, derivatives, and preparative notes are given. (In German)

TABLE 4.1 Aliphatic Aldehydes
(Derivatives preparation, pages 57–60)

Aldehyde	b.p.	2,4-dinitro-phenyl-hydrazone	p-nitro-phenyl-hydrazone	semi-carbazone	dimedone	dimedone anhydride
formaldehyde	−21	166°	182°	169°d	189°	171°
acetaldehyde	20	168	129	163	141	174
propionaldehyde (propanal)	49	155	125	89, 154	156	143
glyoxal	50	328	311	270	186	224
acrolein (propenal)*	52	165	151	171	192	163
isobutyraldehyde (2-methylpropanal)	64	187	131	126	154	144
2-methylpropenal*	73	206	—	198	—	—
butyraldehyde (butanal)	75	123	87	106	142	141
trimethylacetaldehyde	75	209	119	190	—	—
3-methylbutanal	93	123	110	132	155	173
crotonaldehyde* (but-2-enal)	104	190	185	199	183	167
pentanal	104	107	74	—	105	113
2-ethylbutanal	117	130	—	99	102	—
4-methylpentanal	121	99	—	127	—	133
paraldehyde (acetaldehyde trimer)	124	(see acetaldehyde)				
hexanal	131	107	80	106	109	—
tetrahydrofurfural	144	204	—	166	—	—
heptanal	155	108	73	109	103	112
furfural* (furan-2-carbaldehyde)	161	214, 230	152	203	162	164
octanal	170	106	80	101	90	101
nonanal	185	100	—	24, 100	86	—
(+)-citronellal	207	78	—	84	79	173
decanal	208	104	—	102	92	—
aldol (3-hydroxybutanal)	† m.p.	—	—	110	147	126
chloral hydrate	53	131	—	—	56	—

* Electronic absorption spectrum also helpful (chapter 6).
† Decomposes on heating at 760 mm Hg.

TABLE 4.2 Aromatic Aldehydes
(Derivatives preparation, pages 57–60)

Aldehyde	b.p.	m.p.	2,4-dinitrophenyl-hydrazone	p-nitrophenyl-hydrazone	semi-carbazone	oxime	dimedone	dimedone anhydride
benzaldehyde	179°	—	237	192	224	35 d	195	200
phenylacetaldehyde	194	—	121	151	156	99	165	126
salicylaldehyde (o-hydroxybenzaldehyde)	197	—	252	228	231	63	211	208
m-tolualdehyde	199	—	194	157	223	60	172	206
o-tolualdehyde	200	—	194	222	212	49	167	215
p-tolualdehyde	204	—	233	201	234	80	—	—
m-methoxybenzaldehyde	230	—	218	171	233d	40	—	—
p-methoxybenzaldehyde (anisaldehyde)	248	—	254	161	209	65, 132	145	243
cinnamaldehyde (C$_6$H$_5$CH=CHCHO)	252	—	255	195	215	139	213	175
1-naphthaldehyde	—	34	254	234	221	98	—	—
piperonal (3,4-methylenedioxybenzaldehyde)	—	37	265	200	234	110	178	220
o-methoxybenzaldehyde	—	38	253	205	215	92	188	—
2-naphthaldehyde	—	61	270	230	245	156	—	—
vanillin (4-hydroxy-3-methoxybenzaldehyde)	—	81	269	228	239	117	197	228
m-hydroxybenzaldehyde	—	108	259	222	198	90	—	—
p-hydroxybenzaldehyde	—	116	280	266	224	72	189	246
o-nitrobenzaldehyde	—	44	265	263	256	103	—	—
m-nitrobenzaldehyde	—	58	292	247	246	122	—	—
p-nitrobenzaldehyde	—	106	320	249	221	133	—	—
o-chlorobenzaldehyde	213	—	209	249	146, 229	76d	205	225
m-chlorobenzaldehyde	214	—	248	216	229	71d	—	—
p-chlorobenzaldehyde	—	47	265	220	232	107	—	—
o-bromobenzaldehyde	230	22	dec	240	214	102	—	—
m-bromobenzaldehyde	234	—	dec	220	205	72d	—	—
p-bromobenzaldehyde	—	67	dec	208	228	111	—	—

TABLE 4.3 Aliphatic Ketones

(Derivatives preparation, pages 57–60)

Ketone	b.p.	m.p.	2,4-dinitro-phenyl-hydrazone	p-nitro-phenyl-hydrazone	semi-carb.	oxime
acetone	56°	—	128°	149°	190°	59°
ethyl methyl ketone	80	—	115	129	146	—
methyl vinyl ketone‡	80	—	—	—	141	—
biacetyl (butane-2,3-dione)	88	—	315*	230*	279*	234*
isopropyl methyl ketone	94	—	120	109	114	—
methyl n-propyl ketone	102	—	144	117	112	58
diethyl ketone	102	—	156	144	139	69
pinacolone (t-butyl methyl ketone)	106	—	125	—	158	78
isobutyl methyl ketone	117	—	95	79	132	58
di-isopropyl ketone	124	—	88	—	60	34
ethyl n-propyl ketone	124	—	130	—	112	—
n-butyl methyl ketone	128	—	107	88	125	49
mesityl oxide (4-methylpent-3-en-2-one)‡	130	—	203	134	164	49
cyclopentanone	131	—	146	154	210	57
acetylacetone (pentane-2,4-dione)†	139	—	209	—	—	149
di-n-propyl ketone	144	—	75	—	133	—
hydroxyacetone (acetol)	146	—	129	—	196	—
cyclohexanone	156	—	162	147	167	91
hexane-2,4-dione (propionylacetone)†	158	—	—	—	—	—
2-methylcyclohexanone	165	—	137	132	197	43
diacetone alcohol (4-hydroxy-4-methylpentan-2-one)	166	—	203	209	—	58
di-isobutyl ketone	168	—	92	-	122	—
methyl acetoacetate†	170	—	—	224	152	—
3-methylcyclohexanone	170	—	155	119	191	—
4-methylcyclohexanone	171	—	134	128	203	39
cycloheptanone	180	—	148	—	162	—
ethyl acetoacetate†	180	—	93	—	133	—
di-n-butyl ketone	188	—	—	—	90	—
acetonylacetone (hexane-2,5-dione)	194	—	257*	—	220	137*
phorone (2,6-dimethylhepta-2,5-dien-4-one)‡	199	28°	118	—	221	48
(−)-menthone	207	—	142	—	189	59
furoin ⟨furan⟩—COCH—⟨furan⟩ OH		135	217	—-	—	161
furil ⟨furan⟩—COCO—⟨furan⟩	—	165	215	199	—	100*
(+)-camphor	—	179	177	217	238	119
Halogeno ketones						
chloroacetone	119	—	125	—	150	liq.
1,1-dichloroacetone	120	—	—	—	163	—
1,3-dichloroacetone	173	45	133	—	120	—
bromoacetone	140	—	—	—	135	37

* Denotes the double condensation derivative (dioxime, etc.).

† Denotes a compound capable of ready enolization: 2,4-D.N.P. derivative may be a pyrazole (see 4a.1).

‡ Electronic absorption spectrum also helpful (chapter 6).

TABLE 4.4   Aromatic Ketones
(Derivatives preparation, pages 57–60)

Ketone	b.p.	m.p.	2,4-dinitro-phenyl-hydrazone	p-nitro-phenyl-hydrazone	semi-carb.	oxime
acetophenone ($C_6H_5COCH_3$)	202°	—	240°	185°	199°	59°
o-hydroxyacetophenone	215	—	—	—	209	117
o-methylacetophenone	216	—	159	—	203	61
benzyl methyl ketone	216	—	156	145	198	69
propiophenone (PhCOEt)	218	—	191	—	174	53
m-methylacetophenone	220	—	207	—	198	55
isopropyl phenyl ketone	222	—	163	—	181	94
p-methylacetophenone	224	—	258	198	205	88
benzyl ethyl ketone	226	—	—	—	136	—
phenyl n-propyl ketone	230	—	190	—	188	50
m-methoxyacetophenone	240	—	—	—	196	—
n-butyl phenyl ketone	242	—	166	—	166	52
o-methoxyacetophenone	245	—	—	—	183	—
methyl 1-naphthyl ketone	302	34	—	—	229	139
dibenzyl ketone	331	35	100	—	146	125
p-methoxyacetophenone	258	39	220	—	198	87
benzalacetone ($PhCH\!=\!CHCOCH_3$)	262	42	227	166	186	116
benzophenone (diphenyl ketone)	305	48	238	155	165	144
methyl 2-naphthyl ketone	—	56	262	—	236	145
benzalacetophenone (PhCOCH=CHPh)	—	58	245	—	168	115
p-methoxybenzophenone	—	62	180	199	—	138
fluorenone	241	83	284	269	—	195
benzil (PhCOCOPh)	—	95	189	290	244*	237
m-hydroxyacetophenone†	—	96	—	—	195	—
p-hydroxyacetophenone†	—	109	261	—	199	145
benzoin (PhCOCH(OH)Ph)	—	137	245	—	206	151
2,4-dihydroxyacetophenone† (resacetophenone)	—	147	—	—	218	199
m-nitroacetophenone	—	81	228	—	257	132
o-chloroacetophenone	228	228	206	215	160	113
m-chloroacetophenone	228	—	—	176	232	88
p-chloroacetophenone	236	—	231	239	201	95
ω-bromoacetophenone	—	51	220	—	146	90
ω-chloroacetophenone	245	54	215	—	156	89
p-chlorobenzophenone	—	78	185	—	—	156

*Denotes the double condensation derivative (dioxime, etc.).
† These compounds show phenolic properties also (class 2h).

TABLE 4.5   Quinones
(Derivatives preparation, pages 57–60)

Quinone	colour	m.p.	2,4-dinitro-phenyl-hydrazone	p-nitro-phenyl-hydrazone	semi-carb.	oxime	hydro-quinone acetate
p-benzoquinone	yel.	116°	186°(m) 231 (di)	—	166°(m) 178 (m) 243 (di)	240°(di)	123°
1,4-naphthoquinone	yel.	118	278 (m)	279°(m)	247 (m)	—	128
1,2-naphthoquinone	red	147	—	235 (m)	184 (m)	110 (m) 169 (m)	105
9,10-phenanthrene-quinone	or.	208	313 (m)	245 (m)	220 (m)	158 (m)	202
acenaphthenequinone	yel.	261	—	247 (m)	193 (m) 271 (di)	230 (m)	—
9,10-anthraquinone	yel.	286	—	—	—	224 (m)	260
alizarin (1,2-dihydroxy-anthraquinone)	or.	290	—	—	—	—	182
chloranil (2,3,5,6-tetra-chlorobenzoquinone)	yel.	290	—	—	—	—	251

(m) Denotes the mono-condensation product (monoxime, etc.).
(di) Denotes the double condensation derivative (dioxime, etc.).

TABLE 4.6 ## TABLE 4.6 Aliphatic Esters and Lactones
(Derivatives preparation, pages 60–61)

Halogen esters are listed at the end. Esters of carbamic acid are treated as amides (table 4.32).

Ester	b.p.	Ester	b.p.
methyl formate	32°	isobutyl 2-methylpropanoate	149
ethyl formate	54	ethyl lactate (CH$_3$CH(OH)CO$_2$Et)	154
methyl acetate	57	ethyl pyruvate (CH$_3$COCO$_2$Et)	155
isopropyl formate	68	isobutyl butanoate	157
vinyl acetate (CH$_3$COOCH=CH$_2$)	72	isoamyl propanoate	160
ethyl acetate	77	ethyl glycollate (HOCH$_2$CO$_2$Et)	160
methyl propanoate	79	cyclohexyl formate	162
propyl formate	81	butyl butanoate	165
allyl formate (HCOOCH$_2$CH=CH$_2$)	83	isopropyl lactate	168
methyl acrylate (propenoate)	85	(CH$_3$CH(OH)CO$_2$CH(CH$_3$)$_2$)	
dimethyl carbonate (CO(OCH$_3$)$_2$)	90	methyl acetoacetate*	170
isopropyl acetate	91	(CH$_3$COCH$_2$CO$_2$Me)	
methyl 2-methylpropanoate	92	cyclohexyl acetate	175
t-butyl acetate	97	isoamyl butanoate	178
isobutyl formate	98	dimethyl malonate	181
ethyl propanoate	98	ethyl acetoacetate*	
ethyl acrylate (propenoate)	101	(CH$_3$COCH$_2$CO$_2$Et)	181
trimethyl orthoformate	101	ethylene glycol monoacetate	183
(HC(OCH$_3$)$_3$)		diethyl oxalate	186
propyl acetate	101	di-isopropyl oxalate	189
methyl butanoate	102	ethylene glycol diacetate	190
allyl acetate	103	dimethyl succinate	196
(CH$_3$COOCH$_2$CH=CH$_2$)		diethyl malonate	198
butyl formate	107	γ-butyrolactone	204
ethyl 2-methylpropanoate	110	dipropyl oxalate	213
s-butyl acetate	111	diethyl succinate	218
isopropyl propanoate	111	diethyl maleate	223
isobutyl acetate	116	dibutyl oxalate	243
ethyl butanoate	120	dipropyl succinate	246
propyl propanoate	122	glyceryl triacetate (triacetin)	258
isoamyl formate	123	diethyl (+)-tartrate	280
(HCOOCH$_2$CH$_2$CH(CH$_3$)$_2$)		triethyl citrate	294
butyl acetate	125		m.p.
diethyl carbonate (CO(OC$_2$H$_5$)$_2$)	126	ethyl palmitate (C$_{16}$ acid)	24
methyl pentanoate	127	ethyl stearate (C$_{18}$ acid)	33
isopropyl butanoate	128	dimethyl oxalate	53
amyl formate (pentyl formate)	132	dimethyl (+)-tartrate	61
ethyl 3-methylbutanoate	134	trimethyl citrate	78
methyl pyruvate (CH$_3$COCO$_2$Me)	136		
isobutyl propanoate	137		
ethyl but-2-enoate	138	*Chloro esters*	b.p.
isoamyl acetate	142	methyl chloroformate	73
(CH$_3$COOCH$_2$CH$_2$CH(CH$_3$)$_2$)		ethyl chloroformate	93
2-methoxyethyl acetate	144	methyl chloroacetate	129
propyl butanoate	144	ethyl chloroacetate	142
butyl propanoate	145	ethyl 2-chloropropanoate	146
ethyl pentanoate (valerate)	145	methyl trichloroacetate	152
triethyl orthoformate	145	ethyl trichloroacetate	164
(HC(OCH$_2$CH$_3$)$_3$)			
methyl lactate	145	*Bromo esters*	
(CH$_3$CH(OH)CO$_2$Me)		methyl bromoacetate	144
amyl acetate (pentyl acetate)	146	ethyl bromoacetate	159
di-isopropyl carbonate	147	ethyl 2-bromopropanoate	162
(CO(OPr^{iso})$_2$)		ethyl 3-bromopropanoate	179

* See also table 4.21.

Notes: isobutyl = —CH$_2$CH(CH$_3$)$_2$ isoamyl = —CH$_2$CH$_2$CH(CH$_3$)$_2$

s-butyl = —CHCH$_2$CH$_3$ t-butyl = —C(CH$_3$)$_3$
$\quad\quad\quad\quad$ |
$\quad\quad\quad\quad$ CH$_3$

See table 4.17 for structures of acids with common names.

TABLE 4.7   Aromatic Esters and Lactones

(Derivatives preparation, pages 60–61)

Esters of p-nitrobenzoic and 3,5-dinitrobenzoic acids are given in tables 4.9, 4.10, and 4.20.
Aryl benzoates are given in table 4.20.
Esters of 3-nitrophthalic acid are given in tables 4.9 and 4.10.
Esters containing nitrogen or halogen are given at the end, but amino esters are treated as amines (table 4.27).

Ester	b.p.
phenyl acetate	196°
methyl benzoate	198
benzyl formate	203
o-cresyl acetate	208
phenyl propanoate	211
m-cresyl acetate	212
p-cresyl acetate	212
ethyl benzoate	213
methyl o-toluate	213
methyl m-toluate	215
benzyl acetate	216
methyl phenylacetate	216
isopropyl benzoate	218
benzyl propanoate	222
methyl salicylate*	224
ethyl phenylacetate	228
propyl benzoate	230
ethyl salicylate*	234
isopropyl salicylate*	237
benzyl butanoate	238
propyl salicylate*	239
isobutyl benzoate	241
butyl benzoate	248
isoamyl benzoate	262
butyl salicylate*	268
ethyl anisate (p-methoxybenzoate)	269
ethyl benzoylacetate	269
ethyl cinnamate (PhCH=CHCO$_2$Et)	271
resorcinol diacetate	278
dimethyl phthalate	282
diethyl phthalate	298
benzyl phenylacetate	317
benzyl salicylate*	320
benzyl benzoate	323
dibutyl phthalate	339
	m.p.
methyl p-toluate	33
methyl cinnamate (b.p.262)	36
ethyl (±)-mandelate (PhCH(OH)CO$_2$Et)	37
benzyl cinnamate	39
dibenzyl phthalate	42
dibenzyl succinate	42
phenyl salicylate (salol)*	43
cinnamyl cinnamate	44
diethyl terephthalate	44
methyl anisate (p-methoxybenzoate)	45
1-naphthyl acetate	49

Ester	m.p.
methyl (±)-mandelate	58
coumarin (lactone)	67
methyl m-hydroxybenzoate*	70
2-naphthyl acetate	70
ethyl m-hydroxybenzoate*	72
phenyl cinnamate	72
ethylene glycol dibenzoate	73
phthalide (lactone)†	73
2-naphthyl salicylate*	95
ethyl p-hydroxybenzoate*	116
resorcinol dibenzoate	117
diphenyl succinate	121
hydroquinone diacetate	123
methyl p-hydroxybenzoate*	131
dimethyl terephthalate	141
pyrogallol triacetate	165

Esters containing N	b.p.
methyl o-nitrobenzoate	275
	m.p.
ethyl o-nitrobenzoate	30
ethyl m-nitrobenzoate	41
ethyl o-nitrocinnamate	44
methyl o-nitrocinnamate	72
ethyl m-nitrocinnamate	78
methyl m-nitrobenzoate	78
methyl m-nitrocinnamate	123
ethyl p-nitrocinnamate	137
methyl p-nitrocinnamate	161

Chloro esters	b.p.
methyl o-chlorobenzoate	234
ethyl p-chlorobenzoate	238
ethyl o-chlorobenzoate	243
ethyl m-chlorobenzoate	245
	m.p.
methyl m-chlorobenzoate (b.p. 231)	21
methyl p-chlorobenzoate	43

Bromo esters	b.p.
methyl o-bromobenzoate	244
ethyl o-bromobenzoate	254
ethyl m-bromobenzoate	259
ethyl p-bromobenzoate	263
	m.p.
methyl m-bromobenzoate	29
methyl p-bromobenzoate	81

* Show phenolic properties.

†

TABLE 4.8   Carbohydrates
(Derivatives preparation, pages 61–62)

Carbohydrate	$[\alpha]_D^{20}$	acetate m.p.		osazone formation time (min)
		α	β	
D-(−)-fructose	−92°	70°	109°	2
D-(−)-ribose	−22	—	—	—
D-(+)-gentiobiose	+10	188	192	—
D-(+)-mannose	+14	74	115	0·5
D-(+)-cellobiose	+35	228	202	—
D-(+)-glucose	+52	112	134	5
D-(+)-lactose	+53	152	100	appears on cooling
D-(+)-sucrose	+66	70	—	30
D-(+)-galactose	+81	95	142	19
L-(+)-arabinose	+105	—	—	9
D-(+)-maltose	+129	125	158	appears on cooling

TABLE 4.9   Aliphatic Alcohols

(Derivatives preparation, pages 62–64)

Alcohol	b.p.	3,5-di-nitro-benzoate	p-nitro-benzoate	1-naphthyl urethane	hydrogen 3-nitro-phthalate
methanol	65°	109°	96°	124°	153°
ethanol	78	94	57	79	157
propan-2-ol	82	122	110	106	153
2-methylpropan-2-ol (t-butyl alcohol)	83	142	116	101	—
allyl alcohol (CH$_2$=CHCH$_2$OH)	97	50	28	109	124
propan-1-ol	97	75	35	80	145
butan-2-ol (s-butyl alcohol)	99	76	25	98	131
2-methylbutan-2-ol (t-amyl alcohol)	102	118	85	72	—
2-methylpropan-1-ol (isobutyl alcohol)	108	88	68	104	179
3-methylbutan-2-ol	113	76	—	109	127
2,2-dimethylpropan-1-ol (neopentyl alcohol)	113	—	—	100	—
pentan-3-ol	116	100	—	95	121
butan-1-ol (n-butyl alcohol)	118	64	36	72	147
pentan-2-ol	119	62	—	76	103
2-methylbutan-1-ol	129	70	—	82	158
3-methylbutan-1-ol (isoamyl alcohol)	132	62	—	68	166
2-ethoxyethanol	135	76	—	67	121
hexan-3-ol	135	77	—	—	—
pentan-1-ol (n-amyl)	138	46	—	68	136
cyclopentanol	141	115	62	118	—
hexan-1-ol	156	61	—	59	124
cyclohexanol (m.p. 25°)	161	113	50	129	160
furfuryl alcohol	170	81	76	129	—
heptan-1-ol	176	48	—	62	127
propane-1,2-diol (propylene glycol)	189	—	127	—	—
octan-1-ol	194	62	—	66	128
ethane-1,2-diol (ethylene glycol)	198	169	141	176	—
nonan-1-ol	214	52	—	65	125
propane-1,3-diol (trimethylene glycol)	215	178	119	164	—
isoborneol	216	—	138	—	130
geraniol	230	63	35	48	117
decan-1-ol	231	57	30	71	123
diethylene glycol (digol)	244	149	—	122	—
glycerol	290d	—	188	192	—
	m.p.				
dodecan-1-ol (lauryl)	24	60	45	80	124
α-terpineol	35	79	97	152	—
pinacol	38	175	(diacetate m.p. 65°)		
tetradecan-1-ol (myristyl)	39	67	51	82	123
menthol	42	135	62	120	—
hexadecan-1-ol (cetyl)	50	66	52	82	120
but-2-yne-1,4-diol	55	191	—	—	—
octadecan-1-ol (stearyl)	59	66	64	89	119
cholesterol*	148	—	—	160	—
(+)-borneol	208	154	153	127	—
Halogeno alcohols	b.p.				
1-chloropropan-2-ol	127	77	—	—	—
2-chloroethanol (ethylene chlorohydrin)	129	95	56	101	98
2-chloropropan-1-ol	134	75	—	—	—
1-bromopropan-2-ol	148	—	—	—	—
2-bromoethanol (ethylene bromohydrin)	155	—	—	86	175

* $[\alpha]_D^{20}$, $-9\cdot5°$. Acetate m.p. 149°

TABLE 4.10   Aromatic Alcohols
(Derivatives preparation, pages 62–64)

Alcohol	b.p.	3,5-di-nitro-benzoate	p-nitro-benzoate	1-naphthyl urethane	hydrogen 3-nitro-phthalate
1-phenylethanol*	203	94	43	106°	—
benzyl (PhCH₂OH)	205	113	86	134	176°
1-phenylpropan-1-ol*	219	—	60	102	—
2-phenylethanol	220	108	63	119	123
3-phenylpropan-1-ol	237	92	46	—	117
cinnamyl (m.p. 33)	257	121	78	114	—
p-methoxybenzyl (m.p. 25) (anisyl)	259	—	94	—	—
	m.p.				
diphenylmethanol	69	149	131	136	—
benzoin (see table 4.4)	137	—	123	140	—
triphenylmethanol†	162	—	—	—	—

* Readily lose water (on heating with dilute acid) to give styrene or methylstyrene respectively.
† Converted to chloride (m.p. 108) on 10 min. reflux with acetyl chloride and light petroleum: cool, filter off, and protect from moisture.

TABLE 4.11   Aliphatic Ethers
(Derivatives preparation, page 64)

Ether	b.p.
diethyl	35
tetrahydrofuran (T.H.F.)	65
di-isopropyl	68
n-butyl methyl	70
ethylene glycol dimethyl	83
tetrahydropyran	88
di-n-propyl	90
n-butyl ethyl	92
dioxan	101
di-s-butyl	121
ethylene glycol diethyl	123
di-isobutyl	123
di-n-butyl	141
diethylene glycol dimethyl (diglyme)	162
diethylene glycol diethyl	187
di-n-amyl	187

TABLE 4.12   Aromatic Ethers

(Derivatives preparation, pages 64–67)

Ether	b.p.	picrate	nitro derivative	bromo derivative
furan	32°	—	—	—
anisole (PhOMe)	154	81°	87°(2,4-)	61°(2,4-)
phenetole (PhOEt)	170	92	58 (4-)	—
benzyl methyl	171	116	—	—
o-methoxytoluene	171	119	—	—
p-methoxytoluene	175	89	—	—
m-methoxytoluene	177	114	—	—
1,2-epoxyethylbenzene (styrene oxide)	192	—	—	—
o-methoxyphenol (guaiacol)* (m.p. 28°)	205	88	—	116 (4,5,6-)
o-dimethoxybenzene (veratrole)	206	57	—	93 (4,5-)
m-dimethoxybenzene	217	58	72 (2,4-) 157 (4,6-)	140 (4,6-)
diphenyl (m.p. 28°)	259	110	144 (4,4'-)	55 (4,4'-)
methyl 1-naphthyl†	271	129	mixture	55 (2,4-)
ethyl 1-naphthyl†	280	119	mixture	48 (4-)
ethyl 2-naphthyl† (b.p. 282°)	m.p. 37	101	—	66 (1-) 94 (1,6-)
p-dimethoxybenzene (hydroquinone dimethyl)	72	—	49 (2-)	—
methyl 2-naphthyl (Nerolin)†	73	117	128 (1-)	mixture
Halogeno ethers	b.p.			
m-chloroanisole	194	—	—	—
o-chloroanisole	195	—	95	—
p-chloroanisole	198	—	98	—
o-bromoanisole	210	—	106	—
m-bromoanisole	211	—	—	—
p-bromoanisole	215	—	88	—

* Shows phenolic properties.

† Electronic absorption spectrum confirms the naphthalene ring system (chapter 6).

TABLE 4.13   Alkanes and Cycloalkanes

Hydrocarbon	b.p.
2-methylbutane (isobutane)	28
pentane	36
cyclopentane	49
2,2-dimethylbutane	49
2,3-dimethylbutane	58
2-methylpentane	60
hexane	69
methylcyclopentane	72
cyclohexane	81
2-methylhexane	90
3-methylhexane	92
heptane	98
2,2,4-trimethylpentane	99
methylcyclohexane	101
cycloheptane	119
octane	125
ethylcyclohexane	130
methylcycloheptane	134
nonane	151
propylcyclohexane	155
isopropylcyclohexane	155
2,7-dimethyloctane	160
trans-p-menthane	161
cis-p-menthane	169
decane	174
butylcyclohexane	177
trans-decalin*	185
isoamylcyclohexane	193
cis-decalin†	194
undecane	196
amylcyclohexane	200
tetralin‡	207
dodecane	216
tridecane	235
bicyclohexyl	237
tetradecane	254
pentadecane	270 (m.p. 10)
hexadecane	289 (m.p. 18)
octadecane	308 (m.p. 28)

* trans-decahydronaphthalene.
† cis-decahydronaphthalene.
‡ 1,2,3,4-tetrahydronaphthalene.

TABLE 4.14   Alkenes and Cycloalkenes

Hydrocarbon	b.p.
pent-1-ene	30
2-methylbut-1-ene	31
2-methylbutadiene (isoprene)*	34
trans-pent-2-ene	36
cis-pent-2-ene	37
2-methylbut-2-ene (trimethylethylene)	39
cyclopentadiene*†	41
penta-1,3-diene*	42
cyclopentene	44
hexa-1,5-diene	59
hex-1-ene	64
trans-hex-2-ene	68
cis-hex-2-ene	69
2,3-dimethylbutadiene*	69
cyclohexa-1,3-diene*	80
cyclohexene	83
hept-1-ene	94
cycloheptene	114
oct-1-ene	121
cyclo-octatetraene*	141
non-1-ene	147
α-pinene	156
camphene	160 (m.p. 51)
dicyclopentadiene*†	170 (m.p. 32)
dipentene, (±)-limonene	178

* Electronic absorption spectrum also helpful (chapter 6).
† Cyclopentadiene dimerizes rapidly at room temperature by the Diels–Alder reaction to give mainly dicyclopentadiene; the reaction is easily reversed at high temperature and by fractional distillation, regenerating monomeric cyclopentadiene.

monomer   dimer
(conjugated)   (non-conjugated)

TABLE 4.15   Alkynes

Hydrocarbon	b.p.
pent-1-yne	40
pent-2-yne	56
hex-1-yne	71
hex-3-yne	82
hex-2-yne	84
hept-1-yne	100
hept-3-yne	106
hept-2-yne	112
oct-1-yne	126
phenylacetylene	144
non-1-yne	151
dec-1-yne	174

TABLE 4.16 Aromatic Hydrocarbons
(Derivatives preparation, pages 64–67)

In all cases, the electronic absorption spectrum will help to identify the aromatic ring system (chapter 6 and table 6.3).

Hydrocarbon	b.p.	m.p.	picrate	aroyl-benzoic acid	nitro derivative
benzene	80°	—	—	128°	90°(1,3-)
toluene	111	—	—	138	70 (2,4-)
ethylbenzene	135	—	97°	128	37 (2,4,6-)
p-xylene	138	—	—	132	139 (2,3,5-)
m-xylene	139	—	—	126	183 (2,4,6-)
phenylacetylene*	140	—	—	—	—
o-xylene	144	—	—	167	118 (4,5-)
styrene (PhCH=CH₂)	146	—	—	—	—
isopropylbenzene (cumene)	153	—	—	134	109 (2,4,6-)
propylbenzene	159	—	103	126	—
1,3,5-trimethylbenzene (mesitylene)	164	—	97	212	235 (2,4,6-)
t-butylbenzene	169	—	—	—	62 (2,4-)
1,2,4-trimethylbenzene (pseudo-cumene)	169	—	97	149	185 (3,5,6-)
p-isopropyltoluene (p-cymene)	177	—	—	124	54 (2,6-)
indene	182	—	98	—	—
tetralin (1,2,3,4-tetrahydronaphthalene)	207	—	—	154	95 (5,7-)
1-methylnaphthalene	241	—	141	168	71 (4-)
diphenylmethane	262	25°	—	—	172 (2,2′,4,4′-)
2-methylnaphthalene	241	34	115	190	81 (1-)
biphenyl	255	70	—	220	234 (4,4′-)
naphthalene	218	80	150	173	61 (1-)
phenalene	—	85	110	—	—
triphenylmethane	358	92	—	—	212 (4,4′,4″-)
acenaphthylene	—	93	202	—	—
acenaphthene	278	95	162	200	101 (5-)
phenanthrene	340	100	143	—	—
fluoranthene	—	110	185	—	—
fluorene	294	114	84	228	156 (2-)
cis-stilbene	6	150/17	—	—	—
trans-stilbene	306	124	95	—	—
pyrene	—	149	227	—	—
1,1′-binaphthyl	—	160	145	—	—
2,2′-binaphthyl	—	188	184	—	—
anthracene	340	216	138	—	—
chrysene	448	254	273	214	—
perylene	—	274	222	—	—

* Infrared evidence for —C≡CH is helpful (see charts).

TABLE 4.17 Aliphatic Carboxylic Acids
(Derivatives preparation, pages 68–69)

Acid	b.p.	m.p.	S-benzyl-thiuronium salt	anilide	p-tolui-dide	amide	p-bromo-phenacyl ester	p-nitro-benzyl ester
formic	101°	—	151°	50°	53°	—	140°	31°
acetic	118	16°	136	114	153	82°	86	78
acrylic (CH_2=$CHCO_2H$)	140	—	—	105	141	85	—	—
propanoic	141	—	152	106	126	79	63	31
2-methylpropanoic	154	—	149	105	109	129	77	—
butanoic	163	—	149	96	75	115	63	35
pyruvic (CH_3COCO_2H)	165d	13	—	104	130	125	—	—
crotonic (cis)(but-2-enoic)	165	15	—	102	—	102	—	—
3-methylbutanoic	176	—	159	110	109	136	68	—
pentanoic	186	—	156	63	74	106	75	—
methoxyacetic	203	—	—	58	—	96	—	—
hexanoic	205	—	159	—	74	100	72	—
ethoxyacetic	207	—	—	95	—	82	104	—
heptanoic	223	—	—	71	80	96	72	—
cyclohexanecarboxylic (hexahydrobenzoic)	233	31	—	144	—	186	—	—
octanoic	239	16	157	57	70	107	67	—
nonanoic (pelargonic)	254	12	—	57	84	99	69	—
decanoic (capric)	269	31	—	70	78	108	67	—
lactic ($CH_3CH(OH)CO_2H$)	122/15 mm	18	18	59	107	79	113	—
dodecanoic (lauric)	—	43	141	78	87	99	76	—
tetradecanoic (myristic)	—	58	139	84	93	103	81	—
hexadecanoic (palmitic)	—	63	141	91	98	106	86	43
oleic (cis-octadec-9-enoic)	223/10 mm	14	—	41	43	76	45	—
octadecanoic (stearic)	—	70	143	94	102	109	90	—
crotonic (trans) (but-2-enoic)	189	72	172	118	132	160	95	67
glycollic ($HOCH_2CO_2H$)	—	79	146	97	143	120	138	107
glutaric(pentane-1,5-dioic)	—	98	161	224	218	175	137	69
citric (hydrated) (I)	—	100	—	199	189	215	148	102
(+) or (−)-malic (II)	—	101	124	197	207	157	179	124
oxalic (dihydrate) (III)	—	101	198	246	268	419d	242	204
maleic (cis-butenedioic)	—	135	163	187	142	181	168	89
malonic (propane-1,3-dioic)	—	135d	147	225	253	170	—	86
meso-tartaric (see IV)	—	140	—	—	—	190	—	93
adipic (hexane-1,6-dioic)	—	152	163	239	241	220	155	106
(+) or (−)-tartaric (see IV)	—	170	—	264	—	196	216	163
succinic (butane-1,4-dioic)	—	185	154	230	255	260	211	88
(+)-camphoric	—	187	—	226	—	193	—	67
(±)-tartaric (see IV)	—	206	—	—	—	226	—	147
fumaric (trans-butenedioic)	—	286*	176, 195	314	—	226	—	151
Acids containing N or Halogen								
cyanoacetic	—	66	—	198	—	120	—	—
2-chloropropanoic	186	—	—	92	124	80	—	—
dichloroacetic	194	10	178	119	153	97	99	—
2-bromopropanoic	206	25	—	99	125	123	—	—
bromoacetic	—	50	—	130	91	91	—	—
trichloroacetic	—	58	148	95	113	141	—	80
chloroacetic	—	63	160	137	162	120	105	—
iodoacetic	—	83	—	144	—	95	—	—

* In sealed tube.

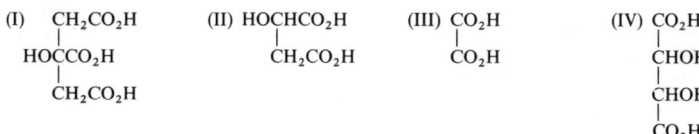

(I) CH_2CO_2H
     |
     $HOCCO_2H$
     |
     CH_2CO_2H

(II) $HOCHCO_2H$
     |
     CH_2CO_2H

(III) CO_2H
      |
      CO_2H

(IV) CO_2H
     |
     $CHOH$
     |
     $CHOH$
     |
     CO_2H

TABLE 4.18   Aromatic Carboxylic Acids
(Derivatives preparation, pages 68–69)

Acid	m.p.	S-benzyl-thiuronium salt	anilide	p-tolui-dide	amide	p-bromo-phenacyl ester	p-nitro-benzyl ester
3-phenylpropanoic (hydrocinnamic)	48°	172°	98°	135°	105°	104°	36°
phenylacetic	76	165	118	136	157	89	65
phenoxyacetic	99	—	101	—	101	148	—
o-methoxybenzoic	101	—	131	—	129	—	113
o-toluic	105	146	125	144	143	57	91
m-methoxybenzoic	110	—	—	—	—	—	—
m-toluic	111	140	126	118	95	108	87
(±)-mandelic (PhCH(OH)CO₂H)	118	166	152	172	134	—	123
benzoic	121	167	162	158	129	119	89
o-benzoylbenzoic†	128	—	195	—	165	—	100
cinnamic (PhCH=CHCO₂H)	133	183	153	168	147	146	117
acetylsalicylic	135	144	136	—	138	—	90
benzilic	150	125	175	190	155	152	100
salicylic (o-hydroxybenzoic)*	159	148	135	156	139	140	98
1-naphthoic	162	—	163	—	202	—	—
p-toluic	178	190	146	*160	159	153	104
p-methoxybenzoic (anisic)	184	185	169	186	162	152	132
2-naphthoic	185	—	170	191	192	—	—
m-hydroxybenzoic*	201	dec 140	157	163	167	176	108
phthalic	ca. 208d	158	251	—	219	153	155
p-hydroxybenzoic*	214	145	197	204	162	191	192
biphenic	229	—	230	—	212	—	186
gallic*	ca. 240d	—	207	—	245	—	—
isophthalic	347	216	—	—	280	179	203
terephthalic	sub. 300	204	337	—	—	225	264
Acids containing N							
m-nitrobenzoic	141	163	154	162	142	132	142
anthranilic (o-aminobenzoic)	146	149	131	151	109	—	205
o-nitrobenzoic	147	159	155	—	175	107	112
4-nitrophthalic	165	180	—	—	200	—	—
m-aminobenzoic	174	—	140	—	111	—	201
2,4-dinitrobenzoic	183	—	—	—	204	158	142
N-benzoylglycine (hippuric)	187	—	208	—	183	151	136
p-aminobenzoic	188	—	—	—	114	—	—
m-nitrocinnamic	205	—	—	—	196	178	174
3,5-dinitrobenzoic	207	—	234	—	183	159	157
3-nitrophthalic	219	219d	234	223	201	—	190
2,4,6-trinitrobenzoic	228	—	—	—	264	—	—
p-nitrobenzoic	239	182	211	203	201	136	169
o-nitrocinnamic	240	—	—	—	185	142	132
p-nitrocinnamic	287	—	—	—	247	191	187
Acids containing halogen							
o-chlorobenzoic	141	177	118	131	141	107	106
o-bromobenzoic	150	171	141	—	155	102	110
m-bromobenzoic	155	168	146	—	155	126	105
m-chlorobenzoic	158	155	124	—	134	117	107
o-iodobenzoic	162	—	141	—	184	110	111
2,4-dichlorobenzoic	164	—	—	—	194	—	—
m-iodobenzoic	187	—	—	—	186	128	121
p-chlorobenzoic	243	223	194	—	179	126	130
p-bromobenzoic	252	—	197	—	189	134	141
p-iodobenzoic	270	—	210	—	218	146	141

* Also give phenolic derivatives: see table 4.20.
† Exists as a lactol: see infrared evidence, and page 65

TABLE 4.19   Carboxylic Acid Anhydrides
(Derivatives preparation, pages 69–70)

Anhydride	b.p.	m.p.
acetic	140	—
propanoic	166	—
butanoic	198	—
maleic	198	56°
glutaric	—	56
succinic	—	120
(+)-camphoric	—	220
o-toluic	—	39
benzoic	360	42
m-toluic	—	71
phenylacetic	—	72
p-toluic	—	95
phthalic	284	132
1,8-naphthalic	—	274
4-nitrophthalic	—	119
3-nitrophthalic	—	164

TABLE 4.20   Phenols
(Derivatives preparation, pages 70–71)

Phenol	b.p.	m.p.	aryloxy-acetic acid	benzoate	p-nitro-benzoate	3,5-dinitro-benzoate	toluene-p-sulphonate	1-naphthyl-urethane	bromo derivative
salicylaldehyde	167	—	132	—	128	—	64	—	—
m-cresol	202	12	103	55	90	165	56	128	84 (tri)
methyl salicylate	223	—	—	92	128	—	—	—	—
ethyl salicylate	231	—	—	80	108	—	—	—	—
m-methoxyphenol	244	28	114	—	—	—	—	129	104
o-methoxyphenol	205	28	119	58	93	142	85	118	116
2,4-dimethylphenol	211	30	142	38	105	164	—	135	179 (tri)
o-cresol	191	36	152	Liq.	94	138	55	142	56 (di)
p-cresol	202	42	136	71	98	189	70	133	49 (di)
phenol	182	43	99	69	126	146	96	—	95 (tri)
phenyl salicylate (salol)	—	49	—	81	111	—	—	—	—
2,6-dimethylphenol	203	51	140	—	99	159	—	160	79
3-hydroxy-4-isopropyltoluene (thymol)	233	56	148	33	70	103	71	—	55
p-methoxyphenol	243	58	111	87	—	—	—	—	—
o-hydroxybiphenyl	275	58	—	76	—	—	65	160	—
3,5-dihydroxytoluene (orcinol) hydrate	289	58	217	88	214	190	—	142	104 (tri)
3,4-dimethylphenol	228	62	163	58	128	181	—	—	171 (tri)
3,5-dimethylphenol	219	68	86	24	109	195	—	—	166 (tri)
2,3-dimethylphenol	218	75	187	—	105	—	—	173	—
2,5-dimethylphenol	211	75	118	61	88	—	—	—	—
4-hydroxy-3-methoxybenzaldehyde (vanillin)	—	81	189	78	—	137	115	—	178 (tri)
o-hydroxybenzyl alcohol (saligenin)	—	87	120	51	—	—	—	—	160
1-naphthol	279	94	192	56	143	217	88	152	105 (di)
t-butylphenol	237	99	86	82	—	—	—	110	—
1,2-dihydroxybenzene (catechol)	240	105	—	84	169	152	—	175	192 (tetra)
3,5-dihydroxytoluene (orcinol) anhydrous	289	108	217	88	214	190	—	160	104 (tri)
m-hydroxybenzaldehyde	240	108	148	38	—	—	—	—	—

Compound	b.p.	m.p.							
1,3-dihydroxybenzene (resorcinol)	280	110	195	117	182	201	81	206	112 (di)
ethyl p-hydroxybenzoate	—	116	—	94	(see p-hydroxybenzoic acid, table 4.18)				—
p-hydroxybenzaldehyde	—	116	198	90	169	210	125	157	84
2-naphthol	285	123	154	107	230	205	—	—	158 (di)
methyl p-hydroxybenzoate	309	131	—	135	(see p-hydroxybenzoic acid, table 4.18)				—
1,2,3-trihydroxybenzene (pyrogallol)	—	133	198	90	205 (picrate, m.p. 96°; triacetate, m.p. 97°)				—
hydroxyhydroquinone (1,2,4-trihydroxybenzene)	—	140	—	120	—	—	—	—	—
salicylic acid	306	159	191	132	250	—	177	—	—
p-hydroxybiphenyl	286	165	—	151	—	—	159	—	186 (di)
hydroquinone	—	170	250	199	—	317	—	247	—
m-hydroxybenzoic acid	—	201	206	70	—	—	—	—	—
p-hydroxybenzoic acid	—	214	278	223	—	—	—	—	—
1,3,5-trihydroxybenzene (phloroglucinol)	—	218	—	174	283	162	—	153	151 (tri)
Halogenophenols									
o-chlorophenol	176	9	145	Liq.	115	143	74	120	95 (tri)
o-bromophenol	195	5	143	71	99	156	78	129	—
m-chlorophenol	214	33	110	86	133	183	53	158	—
m-bromophenol	236	33	108	—	—	—	61	108	—
m-iodophenol	—	40	115	34	—	—	—	—	—
o-iodophenol	217	43	135	86	168	186	71	166	68
p-chlorophenol	210	43	156	96	180	191	125	169	95 (tri)
2,4-dichlorophenol	235	45	140	102	106	136	94	188	—
p-bromophenol	246	64	159	75	153	174	99	153	120 (tetra)
2,4,6-trichlorophenol	—	69	182	119	—	—	—	—	—
p-iodophenol	—	94	156	81	—	—	113	—	—
2,4,6-tribromophenol	—	95	200	—	—	—	—	—	—
Nitrophenols									
o-nitrophenol	216	45	158	59	141	155	83	113	117 (di)
m-nitrophenol	—	97	156	95	174	159	113	167	91 (di)
2,4-dinitrophenol	—	113	—	132	139	—	121	—	118
p-nitrophenol	—	114	187	142	159	186	97	151	142 (di)
2,4,6-trinitrophenol (picric acid)	—	122	—	—	143	—	—	—	—

TABLE 4.21   Aliphatic Enols
(Derivatives preparation, page 70)

Compounds containing easily detectable amounts of the enolic form: b.p. (and m.p.) refer to the equilibrium mixture. For pyrazolones see pp. 58, 74.

Compound	b.p.	m.p.	semi-carbazone	pyrazolone
pentane-2,4-dione (acetylacetone)	139	—	—	—
hexane-2,4-dione (propionylacetone)	158	—	152	127°
methyl acetoacetate	170	—	152	127°
ethyl acetoacetate	180	—	133	127
ethyl acetonedicarboxylate	250d	—	95	85
ethyl oxalacetate	125/17 mm	—	162	—
cyclohexane-1,3-dione*	—	106	—	—

* Monophenylhydrazone, m.p. 177

TABLE 4.22   Aromatic Enols

See note above table 4.21

Compound	b.p.	m.p.	semi-carbazone	pyrazolone
ethyl benzoylacetate	270	—	125	63°
benzoylacetone	262	61	—	63
dibenzoylmethane	—	78	—	137

TABLE 4.23   Primary Aliphatic Amines

(Derivatives preparation, pages 72–73)

Amine	b.p.	benz-amide	toluene-*p*-sulphon-amide	picrate	2,4-dinitro-phenyl derivative
methylamine*	−7°	80°	75°	215°	178°
ethylamine*	17	71	63	165	114
isopropylamine	35	100	51	150	94
t-butylamine	46	134	—	198	—
propylamine	49	84	52	135	96
allylamine	55	—	64	140	76
s-butylamine	63	76	55	140	—
isobutylamine	68	57	78	151	80
butylamine	77	42	65	151	90
amylamine	105	—	—	139	81
ethylenediamine	117	246	160	233	302
hexylamine	129	40	—	127	39
cyclohexylamine	134	149	87	—	156
ethanolamine	171	—	—	160	89
	m.p.				
tetramethylenediamine	28	177°	224	255°d	—
hexamethylenediamine	42	155	—	220	—

* May be encountered as aqueous solutions, which smell similarly to ammonia solutions.

TABLE 4.24 Secondary Aliphatic Amines
(Derivatives preparation, pages 72–73)

Amine	b.p.	benz-amide	toluene-*p*-sulphon-amide	picrate	2,4-dinitro-phenyl derivative
dimethylamine	7	41	79°	158°	87°
diethylamine	56	42	60	155	80
di-isopropylamine	84	—	—	140	—
pyrrolidine	89	—	123	112	—
piperidine	106	48	96	152	93
dipropylamine	110	—	—	75	40
2-methylpiperidine	118	45	55	135	—
3-methylpiperidine	126	—	—	138	67
4-methylpiperidine	128	—	—	—	—
morpholine	130	75	147	148	—
pyrrole	131	—	—	69d	—
dibutylamine	159	—	—	59	—
diamylamine	205	—	—	—	—
	m.p.				
diethanolamine	28	—	99	110	—

TABLE 4.25 Tertiary Aliphatic Amines
(Derivatives preparation, pages 72–73)

Amine	b.p.	methiodide	picrate
trimethylamine	3	230	216°
triethylamine	89	—	173
tripropylamine	156	208	117
tributylamine	212	186	106
triethanolamine	279/150 mm	—	74
	m.p.		
hexamethylenetetramine	280 (sublimes)	190	179

TABLE 4.26 Primary Aromatic Side-Chain Amines
(Derivatives preparation, pages 72–73)

Amine	b.p.	acet-amide	benz-amide	toluene-*p*-sulphon-amide	picrate
benzylamine	185°	60°	105°	116°	196°
(±)-1-phenylethylamine	187	57	120	—	189
2-phenylethylamine	198	114	116	66	174
m-methylbenzylamine	207	240	150	—	198
o-methylbenzylamine	208	69	88	—	215
p-methylbenzylamine	208	108	137	—	204

TABLE 4.27 Primary Aryl Amines and Diamines
(Derivatives preparation, pages 72–73)

Nitro amines are included. Halogeno amines are given at the end.

Warning: regard all aromatic amines as toxic. Many common amines are very toxic (e.g., benzidine, 2-naphthylamine).

HANDLE CAREFULLY. WASH AWAY ACCIDENTAL SPILLAGE WITH SOAP AND WATER.

Amine	b.p.	m.p.	acet-amide	benz-amide	toluene-p-sulphon-amide	picrate	2,4-dinitro-phenyl derivative
aniline	183	—	114	163	103	—	156°
o-toluidine	200	—	112	144	110	213°	126
m-toluidine	203	—	66	125	114	200	161
2-p-xylidine	214	15	142	140	232	—	150
p-ethylaniline	214	—	94	151	—	—	—
2-m-xylidine	215	11	177	168	212	180	—
o-ethylaniline	215	—	112	147	—	—	—
4-m-xylidine	216	—	130	192	181	209	156
5-m-xylidine	220	10	144	136	—	209	—
o-anisidine (o-methoxyaniline)	225	5	88	60	127	200	151
o-phenetidine (o-ethoxyaniline)	228	—	79	104	164	—	164
m-phenetidine	248	—	96	103	157	158	—
m-anisidine	251	—	80	—	68	169	138
p-phenetidine	254	—	135	173	107	69	118
methyl anthranilate*	255	24	101	100	—	106	—
ethyl anthranilate*	266d	13	61	98	112	—	—
ethyl m-aminobenzoate*	294	—	—	114	—	—	—
p-toluidine	200	45	154	158	118	181	137
1-naphthylamine	—	50	160	161	157	163	190
4-aminobiphenyl	—	51	171	230	255	—	—
p-anisidine	—	57	130	154	114	—	141
2-aminopyridine	—	58	71	165	—	221	—
m-phenylenediamine	—	64	191	240	172	184	172
o-nitroaniline	—	71	94	98	110	73	—
4-amino-2-nitrotoluene	—	78	145	172	163	—	—
ethyl p-aminobenzoate*	—	92	110	148	—	131	—
2,4-diaminotoluene	—	99	224	224	192	—	184
o-phenylenediamine	—	102	186	301	202	208	—
2-amino-4-nitrotoluene	—	107	151	186	—	—	—
2-naphthylamine	—	113	134	162	133	195	179
m-nitroaniline	—	114	155	157	139	143	—
benzidine	—	126	317	352	243	—	—
o-tolidine†	—	129	314	265	—	185	—
p-phenylenediamine	—	141	304	300	266	—	177
p-nitroaniline	—	148	216	199	191	100	—
2,4-dinitroaniline	—	180	121	220	219	—	—
2,4,6-trinitroaniline (picramide)	—	190	230	196	—	—	—
4-nitro-1-naphthylamine	—	195	190	224	185	—	—
Halogeno amines							
o-chloroaniline	209	—	88	99	105	134	150
m-chloroaniline	230	—	79	122	138	177	184
m-bromoaniline	251	18	88	120	—	180	—
m-iodoaniline	—	25	119	151	128	—	—
o-bromoaniline	229	32	99	116	90	129	161
o-iodoaniline	—	60	110	139	—	112	—
2,4-dichloroaniline	245	63	146	117	—	106	116
p-iodoaniline	—	63	184	222	—	—	—
p-bromoaniline	—	66	167	204	101	180	158
p-chloroaniline	232	71	179	193	95	178	167
2,4,6-trichloroaniline	263	78	206	174	—	83	—
2,4-dibromoaniline	—	79	146	134	—	124	—
2,4,6-tribromoaniline	—	120	—	232	—	—	—

* Also give chemical and infrared evidence for esters.

† 4.4′-diamino-3,3′-dimethylbiphenyl.

TABLE 4.28   Secondary Aromatic Amines
(Derivatives preparation, pages 72–73)

Amine	b.p.	m.p.	acetamide	benzamide	toluene-p-sulphon-amide	picrate
pyrrole	130°	—	—	—	—	69°d
N-methylbenzylamine	181	—	—	—	95°	—
N-methylaniline	194	—	103°	63°	95	145
N-ethylbenzylamine	199	—	—	—	50	118
N-ethylaniline	205	—	55	60	88	138
N-methyl-m-toluidine	206	—	66	—	—	—
N-methyl-o-toluidine	208	—	56	66	120	90
N-methyl-p-toluidine	210	—	83	53	60	131
N-ethyl-o-toluidine	214	—	—	72	75	—
N-ethyl-p-toluidine	217	—	—	39	71	—
N-ethyl-m-toluidine	221	—	—	72	—	—
dibenzylamine	300d	—	—	112	—	—
tetrahydroisoquinoline	232	—	46	129	—	195
tetrahydroquinoline	250	20°	—	76	—	—
indole	—	52	—	68	—	187
diphenylamine	—	54	103	180	142	182
piperazine	—	104	—	196	—	280
carbazole	—	246	69	98	137	185

TABLE 4.29   Tertiary Aromatic Amines
(Derivatives preparation, pages 72–74)

Amine	b.p.	m.p.	methiodide	picrate	p-nitroso
pyridine	115°	—	118°	167°	—
2-methylpyridine (α-picoline)	129	—	227	169	—
2,6-dimethylpyridine (2,6-lutidine)	142	—	238	163	—
4-methylpyridine (γ-picoline)	143	—	152	167	—
3-methylpyridine (β-picoline)	144	—	92	150	—
2,4-dimethylpyridine (2,4-lutidine)	157	—	113	183	—
N,N-dimethyl-o-toluidine	185	—	210	122	—
N,N-dimethylaniline	193	—	228	164	87°
N-ethyl-N-methylaniline	201	—	125	134	66
methyl nicotinate	204	38°	—	—	—
N,N-diethyl-o-toluidine	210	—	224	180	—
N,N-dimethyl-p-toluidine	211	—	220	130	—
N,N-dimethyl-m-toluidine	212	—	177	131	—
N,N-diethylaniline	215	—	102	142	84
ethyl nicotinate	223	—	—	—	—
N,N-diethyl-p-toluidine	229	—	184	110	—
N,N-diethyl-m-toluidine	231	—	—	97	—
quinoline	238	—	72*(133)	203	—
isoquinoline	242	24	159	223	—
8-hydroxyquinoline	266	76	143d	204	—
triphenylamine	—	127	—	—	—

* Hydrated.

TABLE 4.30   Hydrazines and Semicarbazides
(Derivatives preparation, page 74)

Compound	m.p.	hydrochloride	PhCOMe
phenylhydrazine* (b.p. 243)	19	240	105
p-tolylhydrazine	66	—	—
semicarbazide	96	173d	199
p-nitrophenylhydrazine	157d	—	185
2,4-dinitrophenylhydrazine	194	—	240

* Phenylhydrazine causes dermatitis. Handle with care.

TABLE 4.31   Imines, Schiff's Bases, Aldehyde-Ammonias
(Derivatives preparation, page 74)

Compound	m.p.
benzilidene-p-toluidine	35
benzilideneaniline	54
benzilidene-p-anisidine	62
benzilidene-m-nitroaniline	73
benzilidene-p-phenetidine	78
acetaldehyde-ammonia	93
benzaldehyde-ammonia	102
benzilidene-m-phenylenediamine	105
benzilidene-o-phenylenediamine	106
benzilidene-p-nitroaniline	115
benzilidene-p-phenylenediamine	140

TABLE 4.32   Primary Aliphatic Amides
(Derivatives preparation, pages 74–75)

Amide	m.p.	xanthylamide
ethyl carbamate* (urethane)	49	169
methyl carbamate*	54	193
butyl carbamate*	54	—
isobutyl carbamate*	55	—
propyl carbamate*	61	—
propionamide	79	214
acetamide	82	245
acrylamide	86	—
heptanamide	96	154
pentanamide (n-valeramide)	106	167
butanamide (n-butyramide)	115	187
2-methylpropanamide (isobutyramide)	129	211
urea	132	274
malonamide	170	270
glutaramide	175	—
maleamide	180	—
adipamide	220	—
succinamide	260d	275
oxamide	419d	—

* Also show ester absorptions in the infrared.

TABLE 4.33   Primary Aromatic Amides

(Derivatives preparation, pages 74–75)

Amide	m.p.	xanthylamide
2-phenylpropionamide	92	158
m-toluamide	95	—
3-phenylpropionamide	105	189
benzamide	129	224
o-methoxybenzamide	129	—
(±)-mandelamide	133	—
salicylamide*	139	—
o-toluamide	143	200
cinnamide	148	—
phenylacetamide	157	196
p-toluamide	159	225
p-hydroxybenzamide*	162	—
anisamide (p-methoxybenzamide)	162	—
m-hydroxybenzamide*	167	—
2-naphthamide	192	—
p-nitrobenzamide	201	232
1-naphthamide	202	—
p-ethoxybenzamide	202	—
p-iodobenzamide	218	—
phthalamide	219d	—

* Also show phenolic properties.
For corresponding acids see table 4.18, page 104.

TABLE 4.34   N-Substituted Amides

(Derivatives preparation, pages 75–76)

All *acetamides* ($CH_3CON{\overset{\diagup}{\underset{\big|}{}}}$) and *benzamides* ($PhCON{\overset{\diagup}{\underset{\big|}{}}}$) are listed in tables 4.23 to 4.30 (amine tables).

All *anilides* (RCONPh) and *p-toluidides* ($RCONC_6H_4CH_3$) are listed in tables 4.16 and 4.17 (carboxylic acid tables).

The number of other possible substituted amides is extremely large (and is given by: number of known carboxylic acids multiplied by number of known primary and secondary amines). A few of the commonest are listed below; N-substituted imides are included, since their chemical and spectroscopic properties will lead to their being classified along with tertiary amides in most cases.

Compound	b.p.
N,N-dimethylformamide (D.M.F.)	153
N,N-diethylformamide	176
	m.p.
acetoacetanilide	85
acetoaceto-o-anisidide	87
acetoaceto-p-toluidide	95
acetoaceto-o-toluidide	104
acetoaceto-o-chloroanilide	105
N-phenylsuccinimide	156
N-phenylphthalimide	205
N,N'-diphenylurea (carbanilide)	238

TABLE 4.35 Aminophenols
(Derivatives preparation, pages 72–73, 76)

Aminophenol	m.p.	acetate	benzoate	toluene-p-sulphonate
p-dimethylaminophenol	76	78	158	—
2,4-diaminophenol	79d	222 (di)	231	—
p-methylaminophenol	86	43 (m)	174 (m)	135 (mf)
3-amino-2-hydroxytoluene	89	79 (N)	—	90
o-methylaminophenol	96	64 (di)	160	—
m-aminophenol	123	101 (di)	153 (di)	—
2-amino-6-hydroxytoluene	129	—	—	108
3-amino-4-hydroxytoluene	135	160 (N)	191 (di)	—
3-amino-5-hydroxytoluene	130	—	—	—
2-amino-4-hydroxytoluene	144	129 (di)	—	—
2-amino-3-hydroxytoluene	150	—	189 (N)	—
4-amino-2-hydroxytoluene	161	225 (N)	—	112
4-amino-3-hydroxytoluene	162	171 (N)	162 (di)	—
picramic acid	168	201 (N)	229 (N)	191 (N)
o-aminophenol	174	124 (di)	184 (di)	139
3-amino-6-hydroxytoluene	175	103 (di)	194 (di)	110
2-amino-5-hydroxytoluene	179	130 (N)	92 (m)	—
p-aminophenol	186d	150 (di)	234 (di)	253 (N)
8-amino-2-naphthol	207	165	208	—
1-amino-2-naphthol	dec.	206	235	—

TABLE 4.36 Amino Acids
(α-Amino Acids and Aminobenzoic Acid Isomers)
(Derivatives preparation, pages 76–77)

Physical properties of the (+) form and (±) form may be different: the (−) form has identical properties to the (+) form, except sign of rotation.

Amino Acid	decomp. point (approx.)	benzoate	3,5-dinitro-benzoate	toluene-p-sulphonate
N-phenylglycine	126	63	—	—
anthranilic acid (o-aminobenzoic)	145	182	278	217
m-aminobenzoic acid	174	248	270	—
p-aminobenzoic acid	186	278	290	223
(+)-glutamic acid	198	138	217	117
(+)-lysine	224	150	169	—
(±)-glutamic acid	227	156	—	213
(+)-serine	228	—	—	—
glycine	232	187	179	150
(+)-arginine	238	230	—	—
(±)-serine	246	171	183	213
(+)-cystine	260	181	180	205
(+)-aspartic acid	272	185	—	140
(±)-phenylalanine	274	188	93	135
(±)-tryptophane	275	188	240	176
(±)-aspartic acid	280	165	—	—
(+)-tryptophane	289	104	233	176
(±)-α-alanine	295	166	177	139
(+)-α-alanine	297	151	—	139
(±)-tyrosine	318	197	254	—
(+)-phenylalanine	320	146	93	165
(±)-leucine	332	141	—	—
(+)-leucine	337	107	187	124
(+)-tyrosine	344	166	—	119
(±)-lysine	—	249	—	—

TABLE 4.37   Nitriles
(Derivative preparation, page 77)

Nitrile (cyanide)	b.p.	m.p.
acrylonitrile (vinyl)	78	—
acetonitrile (methyl)	82	—
propionitrile (ethyl)	97	—
isobutyronitrile (isopropyl)	108	—
butyronitrile (propyl)	118	—
but-3-enonitrile (allyl)	118	—
isopentanonitrile (isobutyl)	131	—
pentanonitrile (butyl)	141	—
($\pm$)-mandelonitrile	170d	—
benzonitrile	191	—
malononitrile (methylene di-)	220	31
succinonitrile (ethylene di-)	267	54
methyl cyanoacetate	200	—
o-tolunitrile	205	—
ethyl cyanoacetate	207	—
m-tolunitrile	212	—
phenylacetonitrile	234	—
p-tolunitrile	218	29
1-naphthonitrile	—	36
2-naphthonitrile	—	66
o-nitrobenzonitrile	—	111
m-nitrobenzonitrile	—	118
phthalonitrile	—	141
Nitriles containing Halogen		
m-bromobenzonitrile	—	38
m-chlorobenzonitrile	—	41
o-chlorobenzonitrile	—	43
o-bromobenzonitrile	—	53
p-chlorobenzonitrile	—	96
p-bromobenzonitrile	—	113

TABLE 4.38   Azo Compounds
(Derivative preparation, page 77)

Azo compound	m.p.
2,2'-dimethylazobenzene	55
3,3'-dimethylazobenzene	55
azobenzene	68
3,3'-dimethoxyazobenzene	74
p-dimethylaminoazobenzene	117
p-aminoazobenzene	126
o-hydroxyazobenzene	128
1-benzeneazo-2-naphthol	134
2,2'-dimethoxyazobenzene	147
p-hydroxyazobenzene	152
4,4'-dimethoxyazobenzene	165
4-benzeneazo-1-naphthol	206d

TABLE 4.39   Nitroso Compounds
(Derivatives preparation, page 78)

Nitroso compound	m.p.
nitrosobenzene	68
p-nitroso-N-ethyl-N-methylaniline	68
p-nitroso-N-ethylaniline*	78
p-nitroso-N,N-diethylaniline	84
p-nitroso-N,N-dimethylaniline	87
p-nitroso-N,N-dimethyl-m-toluidine	92
1-nitroso-2-naphthol†	109
p-nitroso-N-methylaniline*	118
p-nitrosophenol†	125d
2-nitroso-1-naphthol†	152d
4-nitroso-1-naphthol†	198

* Isomeric with quinone-imine oximes.
† Exist as quinone monoximes.

TABLE 4.40   Aromatic Nitro Hydrocarbons and Ethers
(Derivatives preparation, page 78)

Compound	b.p.	m.p.
nitrobenzene	211	—
o-nitrotoluene	222	—
2-nitro-m-xylene	226	—
m-nitrotoluene	229	16
2-nitro-p-xylene	237	—
3-nitro-o-xylene	240	15
4-nitro-m-xylene	244	—
o-nitroanisole	265	10
4-nitro-o-xylene	254	30
m-nitroanisole	258	39
p-nitrotoluene	238	54
p-nitroanisole	259	54
p-nitrophenetole	283	60
1-nitronaphthalene	304	61
2,4-dinitrotoluene	—	71
5-nitro-m-xylene	273	74
2-nitronaphthalene	—	79
2,4,6-trinitrotoluene (T.N.T.)	—	82
m-dinitrobenzene	—	90
2,4-dinitroanisole	—	95
o-dinitrobenzene	—	118
1,3,5-trinitrobenzene	—	122
p-dinitrobenzene	—	173
4,4'-dinitrobiphenyl	—	236

TABLE 4.40a   Halogen Substituted Nitro Hydrocarbons and Ethers
(Derivatives preparation, page 78)

Compound	b.p.	m.p.
Chlorides		
o-chloronitrobenzene	245°	33°
m-chloronitrobenzene	—	46
m-nitrobenzyl chloride	—	46
chloro-2,4-dinitrobenzene (2,4-dinitrochlorobenzene)	—	51
p-nitrobenzyl chloride	—	71
chloro-2,4,6-trinitrobenzene (2,4,6-trinitrochlorobenzene, picryl chloride)	—	83
p-chloronitrobenzene	—	83
Bromides		
o-bromonitrobenzene	—	42
o-nitrobenzyl bromide	—	47
m-bromonitrobenzene	—	56
m-nitrobenzyl bromide	—	59
bromo-2,4-dinitrobenzene (2,4-dinitrobromobenzene)	—	75
p-nitrobenzyl bromide	—	100
p-bromonitrobenzene	—	127
Iodides		
m-iodonitrobenzene	—	38
o-iodonitrobenzene	—	54
p-iodonitrobenzene	—	174

TABLE 4.41   Nitro Alkanes

Compound	b.p.
nitromethane	101
nitroethane	114
2-nitropropane	120
1-nitropropane	131

TABLE 4.42 Imides, including Cyclic Urea Derivatives
(Derivatives preparation, pages 78–79)

The barbituric acid derivatives listed are the most important members in current usage. Trivial names are from *British Pharmacopoeia*, 1977, and these differ from American names (ending in *-al*) and from trade names; the electronic absorption spectrum is also helpful (chapter 6).

Imide	m.p.
maleimide	93°
succinimide	125
5-butyl-5-ethylbarbituric acid (butobarbitone)	125
5-ethyl-5-(1-methylbutyl)-barbituric acid (pentobarbitone)	128
5-allyl-5-isopropylbarbituric acid (allobarbitone)	137
5-ethyl-5-isopentylbarbituric acid (amylobarbitone)	168
alloxan (4.H_2O)	170d
5-ethyl-5-cyclohex-1′-enylbarbituric acid (cyclobarbitone)	171
5-ethyl-5-phenylbarbituric acid (phenobarbitone)	177
5,5-diethylbarbituric acid (barbitone)	190
3-nitrophthalimide	216
phthalimide	233
barbituric acid	245d
1,8-naphthalimide	300
5-allyl-5-(1-methylbutyl)-barbituric acid (quinalbarbitone)	—
Imides containing S	
5-ethyl-5-(1-methylbutyl)-2-thiobarbituric acid (thiopentone)	161
o-sulphobenzoic imide (saccharin)	226d
5-allyl-5-cyclohex-1′-enyl-2-thiobarbituric acid (thialbarbitone)	—

TABLE 4.43 Acyl Halides
(Derivatives preparation, page 79)

Acyl halide	b.p.
acetyl chloride	52°
oxalyl chloride	64
methyl chloroformate*	73
propionyl chloride	80
acetyl bromide	81
ethyl chloroformate*	93
butyryl chloride	102
chloroacetyl chloride	105
succinyl chloride	192
benzoyl chloride	197
phthaloyl chloride (di)	281
	m.p.
3,5-dinitrobenzoyl chloride	74
p-nitrobenzoyl chloride	75

* Show ester absorptions in the infrared.

TABLE 4.44   Alkyl Halides and Aromatic Side-Chain Halides

(Derivatives preparation, page 80)

Halide	b.p.	S-alkyl-thiuronium picrate	picrate of 2-naphthyl ether
Chlorides			
ethyl chloride	12	—	102°
isopropyl chloride	37	—	95
methylene chloride	41	267 (di)	—
allyl chloride	45	154	99
propyl chloride	47	177	81
trans-1,2-dichloroethylene	48	—	—
t-butyl chloride	51		—
1,1-dichloroethane	57		—
cis-1,2-dichloroethylene	60		—
chloroform	61	—	—
s-butyl chloride	68		85
isobutyl chloride	69		84
butyl chloride	77	177	67
carbon tetrachloride	77	—	—
1,2-dichloroethane	84		—
trichloroethylene	87	—	
isoamyl chloride	99	173	94
amyl chloride	106	154	67
1,1,2,2-tetrachlorethylene	121	—	—
benzyl chloride	179	188	123
benzal chloride	207	—	—
benzotrichloride	221		—
Bromides			
ethyl bromide	38	188	102
isopropyl bromide	59	196	95
allyl bromide	70		99
propyl bromide	71	177	81
butyl bromide	91	166	85
isobutyl bromide	91	167	84
methylene bromide	97		—
butyl bromide	101	177	67
1,1-dibromoethane	113		—
isoamyl bromide	119	173	94
amyl bromide	129	154	67
1,2-dibromoethane	131		
bromoform	150		
benzyl bromide	198	188	123
Iodides			
methyl iodide	42	224	117
ethyl iodide	73	188	102
isopropyl iodide	89	196	95
allyl iodide	100	154	99
propyl iodide	102	177	81
s-butyl iodide	118	166	85
isobutyl iodide	119	167	84
	m.p.		
iodoform	119	—	—
butyl iodide	129	177	67
isoamyl iodide	147	173	94
amyl iodide	155	154	—

TABLE 4.45   Aryl Halides
(Derivatives preparation, page 81)

Common side-chain halides are included in case of misclassification.

Halide	b.p.	m.p.	nitro derivative
Chlorides			
chlorobenzene	132°	—	52°(2,4-)
			83 (4-)
o-chlorotoluene	159	—	64 (3,5-)
m-chlorotoluene	162	—	91 (4,6-)
p-chlorotoluene	162	7	38 (2-)
benzyl chloride	179	see table 4.44	
o-dichlorobenzene	180	—	110 (4,5-)
benzal chloride	207	see table 4.44	
1,2,4-trichlorobenzene	213	17	56 (5-)
1-chloronaphthalene	259	—	180 (4,5-)
p-dichlorobenzene	—	53	54 (2-)
2-chloronaphthalene	—	61	175 (1,8-)
Bromides			
bromobenzene	156	—	75 (2,4-)
o-bromotoluene	181	—	82 (3,5-)
m-bromotoluene	183	—	103 (4,6-)
p-bromotoluene	185	26	47 (2-)
benzyl bromide	198	see table 4.44	
o-dibromobenzene	224	—	114 (4,5-)
1-bromonaphthalene	281	—	85 (4-)
p-dibromobenzene	—	89	84 (2,5-)
Iodides			
iodobenzene	188	—	174 (4-)
m-iodotoluene	204	—	—
o-iodotoluene	207	—	103 (6-)
p-iodotoluene	211	35	—

TABLE 4.46   Mercaptans, Sulphides, Disulphides and Thioacids
(Derivatives preparation, pages 81–82)

Mercaptan (Thiol)	b.p.	m.p.	2,4-dinitro-phenyl-thioether
Thiols			
methyl	6	—	128°
ethyl	36	—	115
isopropyl	58	—	95
propyl	67	—	81
isobutyl	88	—	76
allyl	90	—	72
butyl	97	—	66
isoamyl	117	—	59
amyl	126	—	80
phenyl (thiophenol)	169	—	121
benzyl	194	—	130
o-thiocresol	194	15	101
m-thiocresol	195	—	91
p-thiocresol	195	44	103
Sulphides			
dimethyl sulphide	38	—	—
methyl sulphide	66	—	—
thiophen	84	—	—
diethyl sulphide	92	—	—
di-isopropyl sulphide	119	—	—
dipropyl sulphide	142	—	—
Disulphides			
dimethyl disulphide	109	—	—
diethyl disulphide	153	—	—
Thioacids			
thioacetic	93	—	—
thiobenzoic	oil	24	—
thiopropionic	liq.	—	—
thio-*p*-toluic	—	44	—

TABLE 4.47 Sulphonic Acids
(Derivatives preparation, pages 82–83)

Arranged in the order of increasing m.p. of the S-benzylthiuronium salts.

Acid	S-benzyl-thiuronium salt	sulphonyl chloride	sulphon-amide	sulphonanilide
1-naphthol-4-sulphonic	104	—	—	200
m-sulphobenzoic*	133	20	170	—
2-naphthol-1-sulphonic	136	124	—	161, 105($2H_2O$)
naphthalene-1-sulphonic	137	68	150	112, 157
benzenesulphonic	150	15	153	110
phenol-p-sulphonic	169	—	177	141
toluene-o-sulphonic	170	10	156	136
1-naphthol-2-sulphonic	170	—	—	—
toluene-m-sulphonic	—	12	108	96
toluene-p-sulphonic	182	71	137	103
naphthalene-2-sulphonic	191	79	217	132
anthraquinone-1-sulphonic	191	217	—	216
o-sulphobenzoic*	206	40, 79	—	195
benzene-o-disulphonic	206	143	254	241
(+)-camphor-10-sulphonic	210	68	132	120
naphthalene-2,7-disulphonic	211	159	243	—
anthraquinone-2-sulphonic	211	197	261	193
p-sulphobenzoic*	213	57	236	252
benzene-m-disulphonic	214	63	229	144
2-naphthol-6-sulphonic	217	—	238	161
benzene-p-disulphonic	—	141	288	249
naphthalene-1,4-disulphonic	—	160	273	179
naphthalene-1,6-disulphonic	235	129	297	—
naphthalene-2,6-disulphonic	256	225	305	—
Acids containing N (aminosulphonic acids, table 4.47a)				
o-nitrobenzenesulphonic	—	69	193	115
m-nitrobenzenesulphonic	146	64	168	126
p-nitrobenzenesulphonic	—	80	179	136
Acids containing Halogen				
o-chlorobenzenesulphonic	—	28	188	—
m-chlorobenzenesulphonic	—	Liq.	148	—
p-chlorobenzenesulphonic	175	53	144	104
o-bromobenzenesulphonic	—	51	186	—
m-bromobenzenesulphonic	—	Liq.	154	—
p-bromobenzenesulphonic	170	75	166	119

* Give double S-benzylthiuronium salts, and dichlorides, diamides, and dianilides.

TABLE 4.47a   Aminosulphonic Acids
(Derivatives preparation, page 85)

Acid	S-benzyl-thiuronium salt
o-aminobenzenesulphonic (orthanilic)	132
m-aminobenzenesulphonic (metanilic)	148
p-aminobenzenesulphonic (sulphanilic)	187
1-naphthylamine-4-sulphonic	195
1-naphthylamine-5-sulphonic	180
1-naphthylamine-6-sulphonic	191
1-naphthylamine-7-sulphonic	—
1-naphthylamine-8-sulphonic	300
2-naphthylamine-1-sulphonic	139
2-naphthylamine-6-sulphonic	184

TABLE 4.48   Sulphonate Esters
(Derivatives preparation, page 83)

The toluene-p-sulphonate esters of many phenols are listed in table 4.20.

Ester	b.p.	m.p.
ethyl methanesulphonate	86°/10 mm	—
methyl benzenesulphonate	150/15 mm	—
ethyl benzenesulphonate	156/15 mm	—
propyl benzenesulphonate	163/15 mm	—
propyl toluene-p-sulphonate	165/10 mm	—
methyl ethanesulphonate	201	—
methyl methanesulphonate	206	—
ethyl ethanesulphonate	214	—
methyl toluene-p-sulphonate	—	28
ethyl toluene-p-sulphonate	—	33
phenyl toluene-o-sulphonate	—	52

TABLE 4.49   Sulphate Esters
(Derivatives preparation, page 83)

Regard all sulphate esters as highly toxic.

Ester	b.p.
dimethyl sulphate	188
diethyl sulphate	208
dipropyl sulphate	94/5 mm

TABLE 4.50   N-Halogeno Compounds

Compound	m.p.
N-chlorosuccinimide	150
N-bromosuccinimide	174d
N-chlorophthalimide	185
N-bromophthalimide	207

TABLE 4.51   Thioamides and Thioureas
(Derivatives preparation, page 84)

Compound	m.p.
thiopropionamide	43
thiopropionanilide	68
thioacetanilide	76
thiobenzanilide	102
thioacetamide	115
thiobenzamide	116
p-methylthiobenzanilide	141
N-phenylthiourea	154
N,N'-diphenylthiourea (thiocarbanilide)	154
p-methylthiobenzamide	168
thiourea	180
thiosemicarbazide	182

TABLE 4.52   N-Halogeno Sulphonamides
(Derivatives preparation, page 84)

Compound	m.p.
N-chlorotoluene-p-sulphonamide (Na salt) (Chloramine-T)	indef.
N,N-dichlorotoluene-p-sulphonamide (Dichloramine-T)	81
N-chlorobenzenesulphonamide (Na salt) (Chloramine-B)	indef.

TABLE 4.53   Hydrohalides of Thioureas
(Derivatives preparation, page 84)

Compound	m.p.
N-methylthiuronium iodide	< 100
thiuronium chloride (thiourea hydrochloride)	< 100
S-benzylthiuronium chloride	175
S-(p-chlorobenzyl)-thiuronium chloride	197

TABLE 4.54   Phosphate and Phosphite Esters

Phosphate ester	b.p.	m.p.
trimethyl phosphate	197	—
triethyl phosphate	216	—
tri-isopropyl phosphate	84/5 mm	—
tripropyl phosphate	108/5 mm	—
tributyl phosphate	139/6 mm	—
triphenyl phosphate	—	50
Phosphite ester		
trimethyl phosphite	112	—
triethyl phosphite	157	—
tripropyl phosphite	207	—
tributyl phosphite	120/10 mm	—
triphenyl phosphite	—	24

5

The Nuclear Magnetic Resonance Spectrum

Nuclear magnetic resonance (n.m.r.) spectroscopy has done nothing less than revolutionize the investigation of structures in organic chemistry. An immense amount of detailed information can be derived from the use of the technique, but the following points are intentionally restricted to the practical applications of only one aspect of n.m.r. spectroscopy—its use in qualitative organic analysis.

Within this context, much information can be obtained from a study of the n.m.r. spectrum by itself. But since infrared analysis and chemical tests have already led to a number of conclusions about the structure of the unknown, these can both simplify the analysis of the n.m.r. spectrum and enable the investigator to draw more definite conclusions than might otherwise be possible.

It is assumed that the student has had an introductory course on n.m.r. spectroscopy, and that the n.m.r. spectrum has been recorded by, or supplied to him with an integral trace recorded on it.

Suitable student texts to complement these specialized notes are the books by Williams and Fleming, by Dyer, by Schwarz, and by Kemp referred to on page 39. The following deal more fully with n.m.r. only:

L. M. Jackman, *Applications of Nuclear Magnetic Resonance Spectroscopy in Organic Chemistry*, Pergamon Press, London, 1959.

J. D. Roberts, *Nuclear Magnetic Resonance*, McGraw-Hill Book Co. Inc., New York, 1959.

Access to a catalogue of reference spectra is also helpful, so that the student can become familiar with the appearance of typical spin-spin splitting patterns (some of which are shown on fig. 5.1). Catalogues of spectra have been published, for example:

NMR Spectra Catalog, Volumes 1 and 2, Varian Associates, Palo Alto, California; or the more comprehensive *Sadtler Research Laboratories Nuclear Magnetic Resonance Spectra*, Heyden and Son, London, 1966.

The three parameters which are extracted from a n.m.r. spectrum are

(a) the chemical shift positions, (b) the coupling constants, and (c) the integrals. All figures given here for chemical shift data are in δ units; students more accustomed to using τ units can quickly convert to these ($\tau = 10 - \delta$). Coupling constant data are given in Hz.

Although it is difficult to lay down a fixed set of rules for examining a n.m.r. spectrum, the student should study each spectrum in a systematic manner. To this end, in all cases, follow through the four steps outlined below in the order given.

5.1   Aromatic, Alkene, and Alkane Protons

The spectrum should first be scanned to decide whether the compound contains aromatic, alkene, or alkane groupings. (Alkyne protons cannot be definitely identified in the first survey, and are detected much more easily in the infrared spectrum: see infrared charts, 1c).

The approximate ranges within which these groups of protons come to resonance are:

> *Aromatic protons:* 6·5–8·5 δ (Benzene appears at 7·27 δ, heterocylic aromatics appear at 6·0–9·0 δ). See page 131.
> *Alkene protons:* 4·5–6·5 δ (Aryl conjugation raises, even as far as 7·8 δ). See page 132.
> *Alkane protons:* 0·7–4·5 δ (Values outside these limits are uncommon). See pages 127–131.
> *Alkyne protons:* See page 133.

It is important to compare these deductions with conclusions already reached from chemical and infrared evidence; the detection of aromatic protons in particular is simple from n.m.r. evidence.

5.2   The Integral Trace

Having made a tentative classification of the protons present in the compound, carefully calculate the relative numbers of protons in each environment from the integral trace.

Divide the spectrum as far as is possible into individual groups of protons; in the case of complex multiplets it will be advisable at this stage to calculate the *total* integral for each multiplet; more detailed calculations can be carried out after the examination of the splitting patterns.

Measure the height of each integral step to the nearest 0·5 mm; thereafter calculate the relative intensities of the peaks, and convert the resulting ratios to integer ratios. Do not 'round off' these ratios with abandon; if more than one set of integer ratios appear to satisfy the peak intensities, then *each* set of ratios should be considered in turn to decide which is correct. Additional help will be obtained from studying the spin splitting patterns, since many characteristic patterns have known integral ratios; see below.

5.3 Singlet Signals

Identify on the spectrum all the singlet peaks, and note whether the signal is sharp, broad, or very broad. All the singlet signals which arise from protons linked to carbon, (a) to (g), are sharp and relatively unaffected by changes in concentration or temperature: solvent shifts are also usually small.

Signals from protons linked to oxygen or nitrogen, (h) to (p), may be sharp or broad; chemical shift positions change with concentration and temperature. They are all moved downfield by hydrogen bonding, particularly signals for OH protons, and the values quoted apply in general to *ca.* 10 per cent solutions in inert solvents; dilution decreases hydrogen bonding (not in carboxylic acids) and moves the signals upfield.

In the case of OH and NH protons, deuteration removes the signal from the spectrum, and the addition of traces of acid (e.g., trifluoroacetic acid) or base (e.g., pyridine) may also cause shifts as indicated below.

For each singlet, consider the following possibilities; while not an exhaustive list, the examples given are the commonest singlet peaks. This survey should be conducted in conjunction with any previously acquired chemical and infrared evidence.

(a) **1·1–3·8 δ. Methyl.** A methyl group will appear as a sharp singlet, integral 3, if it is attached directly to one of the functions listed in table 5.1, from —CO_2R down, excluding CHO. The chemical shift positions listed in table 5.1 will usually suffice to identify such methyl groups. (Methyl groups attached to alkene chains may undergo spin splitting. See 5.4 below.)

(b) **1·0–2·0 δ. t-Butyl.** A sharp singlet, integral 9, invariably indicates the t-butyl group, but three methyl groups in the molecule in identical magnetic environment, for example in mesitylene, will also appear thus. Molecular symmetry of this order is relatively rare compared with the incidence of the t-butyl group.

(c) **2·5–6·0 δ. Methylene.** An isolated methylene group flanked, for example, by two of the groups listed in table 5.1 will appear as a singlet, integral 2, unless restricted rotation holds the two geminal protons in different environments. To predict the chemical shift positions of these protons is difficult, but J. N. Shoolery has computed the relationship shown on table 5.2, which is usually fairly accurate.

Two adjacent methylene groups with identical substituents, as in succinic acid, will also appear as a singlet, but with integral 4. These will appear within the narrower range 2·1–3·5 δ. Restricted rotation about the CH_2—CH_2 bond will complicate this, as the protons may no longer be in identical magnetic environments.

(d) **3·5–7·3** δ. **Methine.** The difficulties encountered with isolated methylene groups arise again with isolated methine protons. The integral will be 1; the expected chemical shift position should be calculated by using table 5.2. The highest values (e.g., 7·3 δ) only arise when all three groups attached to the CH are powerfully deshielding.

(e) **4·5–7·8** δ. **Alkene.** Singlet alkene peaks can usually be identified from the chemical shift positions given in table 5.3. Chemical and infrared evidence must support these assignments. If no aromatic ring is present, the range is narrower (4·5–6·8 δ).

(f) **9·5–10·5** δ. **Aldehyde.** Aryl CHO comes to resonance within the range 9·85–10·0 δ. *Ortho*-nitro groups, etc., raise the signal position to above 10·0 δ. *Formate esters*, etc., which contain the CHO group, show singlet absorption around 8·1 δ.

Aliphatic aldehyde CHO will appear as a multiplet unless the CHO group is attached to a hydrogen-free carbon atom; the signal (around 9·65 δ) is easily assigned, as few other signals arise in this region.

(g) **1·0–4·0** δ. **Alcohols.** On addition of a trace of acid, this fairly sharp singlet moves to near 4·7 δ. At high concentrations the signal will move downfield. Very pure samples only will show coupling with neighbouring protons. Alcohols in dimethyl sulphoxide solution may also show OH coupling. Signal disappears on deuteration.

(h) **1·0–2·0** δ. **Thiols (aliphatic).** These show similar shifts to alcohols.

(i) **4·5–6·0** δ. **Phenols.** These show similar shifts to alcohols. Chelation by *ortho*-substituents will shift the signal to lower field (8–11 δ). Signal disappears on deuteration.

(j) **around 3·5** δ. **Thiophenols.** These show similar shifts to phenols.

(k) **1·5–4·0** δ. **Aliphatic Amines.** Primary amines have an integral of 2, while secondary amines have 1; where more than one amino group is present, the integrals will vary accordingly. On addition of a *trace* of acid the signal moves towards 5 δ. Signal disappears on deuteration.

(l) **3·5–5·0** δ. **Aromatic Amines.** These show similar shifts to aliphatic amines. Signal disappears on deuteration.

(m) **7–10** δ. **Amine Salts.** In trifluoroacetic acid as solvent, amine salts show a singlet signal whose position varies considerably (aliphatic 7–8 δ, aromatic 8–10 δ). In a large excess of acid, the singlet may give way to a very broad triplet. In D_2O, another convenient solvent, the NH signals vanish.

(n) **5·0–8·5** δ. **Amides.** Primary amide protons show an extremely broad signal, which may be lost in the noise level of the spectrum; secondary amide protons usually give a sharper singlet. (N-phenyl amides show a moderately broad singlet around 8 δ.) Imides also show broad singlet peaks, but at lower field (9–12 δ). Signal disappears on deuteration.

(o) **10–13 δ. Carboxylic Acids.** These give a highly characteristic signal above 10 δ. Note that dilution does not shift the carboxyl signal to higher field, since the hydrogen bonding (dimeric association) does not change on dilution. Signal disappears on deuteration.

(p) **11–16 δ. Enols.** Strongly enolic compounds are usually stabilized in this form by strong hydrogen bonding (e.g., acetylacetone), and this accounts for the exceptionally low signal. Less highly enolic compounds (e.g., β-keto esters) will show the signals for the enol *and* the keto forms; in such cases the integrals for the enol protons will not have an integer value, unless separate measurements are made for the enol and keto forms. Such measurements also indicate the keto-enol ratio. Signal disappears on deuteration.

5.4   Simple Spin Splitting Systems

The multiplicity of the signals other than singlets should now be examined. Use a pair of dividers to measure the separation among the various lines; where the signals are undistorted and the multiplicity is relatively simple, then these splittings will correspond to the true coupling constants. *Otherwise this does not hold*, and the true coupling constants can only be obtained by calculation. Coupling constants can be positive or negative, depending on the relative energies of the spin states involved; since the sign of J values cannot be directly extracted from the spectra, the signs are not shown in table 5.4.

Mark those multiplets in which the same splittings can be seen; if the same coupling constant is shown in two separate multiplets, and nowhere else, then these two multiplets represent coupling protons which are probably close together in the molecule. Consult table 5.4, and try to relate the dimensions of the coupling constants with groupings in the molecule.

Look specifically for the following very characteristic sets of multiplets, which occur in a wide range of compounds. The typical appearances of these systems are shown diagrammatically in fig. 5.1. For all undistorted multiplets, the $(n + 1)$ rule should be applied: (multiplicity $= n + 1$, where $n =$ number of coupling neighbours); remember that this first-order process will be of no value in complex multiplets, when the number of lines observed may be many more than $n + 1$.

Unlike infrared analysis, every signal in the n.m.r. spectrum must be assigned to a proton or group of protons in the molecule, and these assignments must be reasonable in terms of chemical shifts, integrals, and coupling constants.

(a) **AX or AB Sysyems.** True AX systems are relatively rare in all but the simplest of molecules, but these show as two undistorted doublets of equal

(a)

AB system

(b)

Ethyl
$\overline{CH_3CH_2}$

(c)

Isopropyl
$\dfrac{CH_3\diagdown}{CH_3\diagup}CH-$

(d)

n-Propyl
$CH_3CH_2CH_2-$
See text for
$CH_3CH_2CH\diagup$

1,3-Propylidene
$-CH_2CH_2CH_2-$

(e)

Aromatic systems

Strongly deshielding (or shielding) substituent

Symmetrical o-disubstitution

Unsymmetrical p-disubstitution

(f)

Alkene systems
Vinyl, AMX
$CH_2=CH-$

Allyl
$CH_2=CHCH_2-$

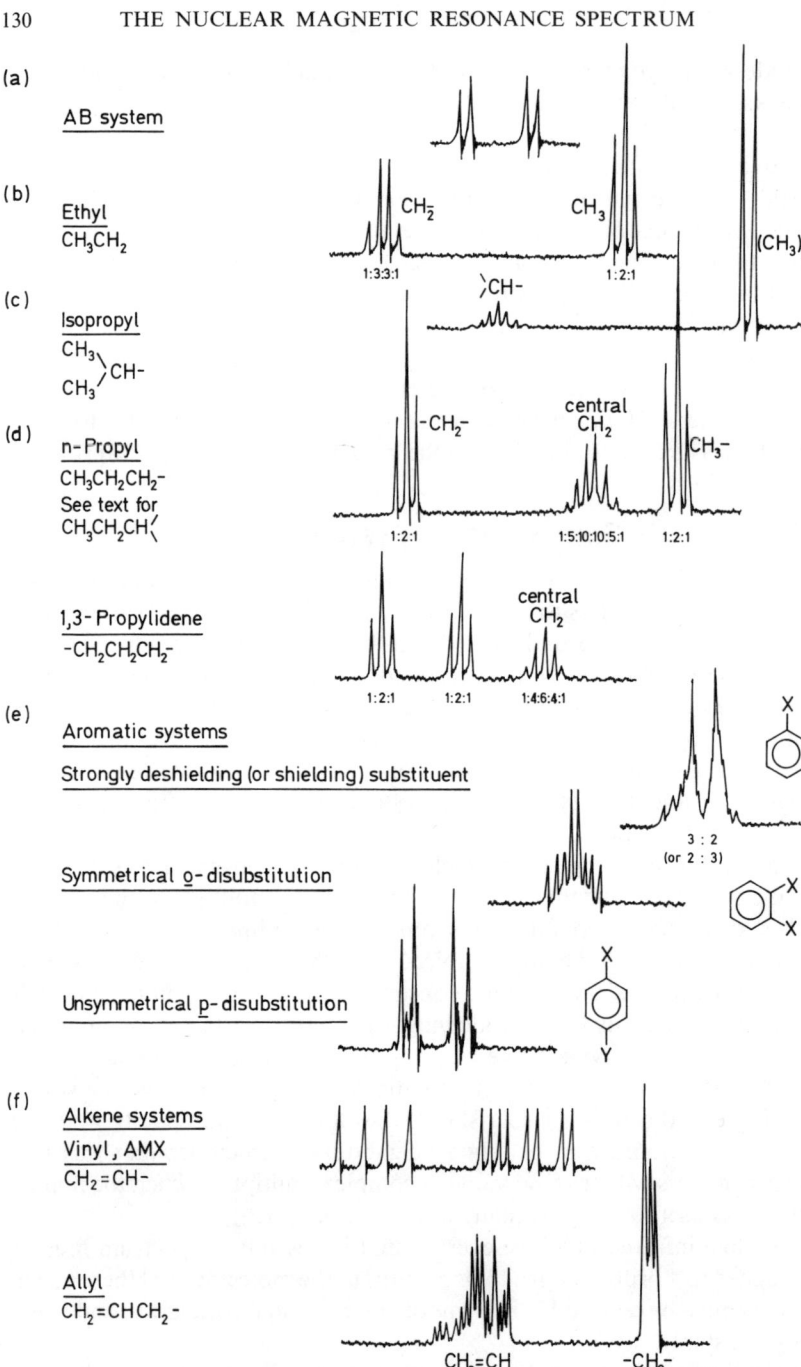

Fig. 5.1   Simple spin splitting systems

intensity with the same coupling constant. For an undistorted AX system to arise, the difference in chemical shift values of the two coupling protons must be at least six times the value of J. More common is the AB case, when the chemical shifts of the two protons are closer; the peaks become distorted, the inner peaks being enlarged while the outer peaks are diminished in intensity. At the same time, the origin position of the chemical shift for each proton is no longer the mid-point of each doublet. The origin chemical shift position can be calculated exactly, but is approximately the 'centre of gravity' of each doublet.

Alkene systems (e.g., an isolated $C=CH_2$, or the $-CH=CH-$ group of cinnamic acid derivatives, etc.) are a frequent source of AB quartets.

Check the chemical shifts of these systems against tables 5.1 and 5.3; coupling constants should be checked against table 5.4.

Unsymmetrical *para*-disubstituted benzene derivatives show signals of similar appearance (see fig. 5.1(e)), but these are discussed under aromatic systems (e).

(b) **Ethyl.** The signal arising from the ethyl group is one of the simplest systems to identify on the n.m.r. spectrum; it consists of a triplet, integral 3, around $1-2\,\delta$, and a quartet, integral 2, at lower field depending on substituent. J is 8–10 Hz; see fig. 5.1(b), and note the relative intensities of the lines within each multiplet. Two ethyl groups may be present in the same magnetic environment (as in ethyl malonate), in which case the integrals will be 6 and 4 respectively.

Use table 5.1 to identify the group attached to CH_2; this must be in accord with chemical and infrared evidence.

(c) **Isopropyl.** This gives rise to a doublet and a septet; the doublet is around $1-2\,\delta$, integral 6, and the septet at lower field, integral 1. All seven lines of the septet may not be seen. J is 8–10 Hz. Use table 5.1 to identify the group attached to CH.

(d) **Propyl and Propylidene.** The groups $CH_3CH_2CH_2-$ and $CH_3CH_2CH\diagdown$ show very similar signals (see fig. 5.1). Both show a triplet for CH_3 around $1\,\delta$; both show a multiplet (sextet and quintet respectively) for the central CH_2 around $2\,\delta$; and both show a second triplet (integrals 2 and 1 for $-CH_2-$ and $-CH\diagdown$ respectively) whose position can be predicted from tables 5.1 or 5.2. The group $-CH_2CH_2CH_2-$ also shows a quintet around $1\cdot3-2\cdot4\,\delta$ for the central CH_2, and two triplets for the outer CH_2 groups; the positions of these triplets can be calculated from table 5.1.

(e) **Aromatic Systems.** If the aromatic protons give rise to a complex multiplet with no apparent symmetry or with no clear separation between groups of signals, then you would be advised simply to note the approximate limits of the chemical shift range, and use these to help in deducing the nature of substituents present (table 5.5). First-order considerations cannot be applied in such circumstances.

The δ values for naphthalene derivatives and other polynuclear aromatic derivatives are higher (by *ca.* 0·4 δ) than for the corresponding benzene derivatives. Ring protons adjacent to nitrogen in heterocyclics such as pyridine, which are also aromatic, are at high δ (see table 5.3); note also that a nuclear proton flanked by two powerful shielding (e.g., NH_2) or deshielding (e.g., NO_2) groups will lie outside the range 6·5–8·5 δ.

Where complex multiplets are shown, but with the peaks separated into two groups (see fig. 5.1(e)), calculate the integrals of the two groups separately, as this indicates that a strongly electronegative or electropositive substituent is present on the nucleus, and the signals for protons *ortho* to this substituent are being moved substantially away from the remaining protons on the ring.

Calculate the integrals of aromatic compounds with particular care, because this enables you to count the number of protons on the ring, and hence the number of substituents present. In this respect, n.m.r. is normally of much greater value than infrared spectroscopy, where deductions about substitution pattern must be made with caution (see page 10).

A complex multiplet with considerable symmetry about the mid-point of the system (see fig. 5.1(e)), is often characteristic of benzene derivatives with two identical substituents in the *ortho*-positions. Chemical shift data will help to identify the substituent (table 5.5).

A highly characteristic set of peaks arises in the majority of *para*-disubstituted benzenes, where the two substituents are of widely differing electronegativities (see fig. 5.1(e)). Superficially, the peaks resemble an AB system (or more accurately an A_2B_2 system), but this is not exactly so, since each proton couples with the proton *ortho* to it and with the proton *meta* to it. These extra couplings give rise to a number of extra lines, particularly within each doublet; first-order analysis is not completely accurate, but if the chemical shift and coupling constant data are extracted as for first-order spectra, they will be accurate enough for identification purposes. Some 1,2,4-trisubstituted benzene derivatives give similar spectra, but the integrals of the two 'doublets' are in the ratio 2:1. Many tetrasubstituted benzene derivatives will show true AB spectra if the chemical shift positions of the two nuclear protons are sufficiently separated; in these cases it is almost always possible to decide whether these protons are *ortho*, *meta*, or *para* to each other from the coupling constants given in table 5.4.

(f) **Alkene Systems.** Alkene groupings which give rise to AB quartets have already been mentioned in (a) above.

Two other commonly encountered multiplet systems arise with vinyl and allyl groups.

Vinyl groups (CH_2=CH—) attached to strongly deshielding substituents (especially oxygen functions) may show on the spectrum as AMX or ABX systems (see fig. 5.1). If the chemical shift positions are closer than *ca.* 2 δ units, however, they will give rise to ABC systems, and no attempt should be made to extract coupling constant data directly; the integral values and the approximate chemical shift positions should be used to identify the protons present.

Allyl groups (CH_2=CHCH$_2$—) produce extremely complex multiplets (see fig. 5.1), but this makes them fairly easy to recognize on the spectrum. The lower multiplets occupy most of the spectrum between 4·8–6·2 δ (overall integral 3). The position of the upper multiplet (integral 2) can be calculated with reasonable certainty from table 5.2.

The *propenyl group* (CH_3CH=CH—) has similar overall appearance, but the upper multiplet (integral 3) always appears around 2 δ, while the lower multiplets vary according to substituent (*ca.* 5–8 δ, see table 5.3).

In all alkene systems, considerable assistance in assignment can be obtained from the coupling constant data given in table 5.4. Where the molecule can exist in geometrically isomeric forms, it is frequently possible to distinguish *cis* from *trans* isomers because of the differing coupling constants; a mixture of *cis* and *trans* forms will show signals appropriate to both.

(g) **Alkyne.** This sharp signal will appear within the range 1·8–3·3 δ (integral 1); it will appear as a singlet only if the HC≡C— group is attached to a hydrogen free atom (e.g., in PhC≡CH). Supporting evidence from infrared spectroscopy is essential, and quite definitive (see page 11). Alkyne proton signals disappear on deuteration, and move to lower field on the addition of base (e.g., pyridine).

TABLE 5.1 δ Values for the Protons of CH_3, CH_2 and CH Groups Attached to groups X, where R = alkyl, and Ar = aryl

X	CH_3X	$R'CH_2X$	$R'R''CHX$
—R	0·9	1·3	1·5
—CH^b $\overset{a}{CH_2}$ O (ring)	1·3	a 3·5	b 3·0
$\diagdown$ =	1·7	1·9	2·6
=—=—=, etc. (i.e., end-of-chain)	1·8		
=—=—=, etc. (i.e., in-chain)	2·0	2·2	2·3
=N—	2·0	—	—
—≡	2·0	2·2	—
—COOR, —COOAr	2·0	2·1	2·2
—CN	2·0	2·5	2·7
—CONH$_2$, —CONR$_2$	2·0	2·0	2·1
—COOH	2·1	2·3	2·6
—COR	2·1	2·4	2·5
—SH, —SR	2·1	2·4	2·5
—NH$_2$, —NR$_2$	2·1	2·5	2·9
—I	2·2	3·1	4·2
—CHO	2·2	2·2	2·4
—Ph	2·3	2·6	2·9
—Br	2·6	3·3	4·1
—NHCOR, —NRCOR	2·9	3·3	3·5
—Cl	3·0	3·4	4·0
—OR	3·3	3·3	3·8
—NR$_3^+$	3·3	3·4	3·5
—OH	3·4	3·6	3·8
—OCOR	3·6	4·1	5·0
—OAr	3·7	3·9	4·0
—OCOAr	3·9	4·2	5·1
—NO$_2$	4·3	4·4	4·6

TABLES 5.1–5.5: Adapted with permission from W. Kemp *Organic Spectroscopy*. Macmillan Press, London, 1975.

TABLE 5.2 δ Values for the Protons of CH_2 (and CH) Groups bearing more than one Functional Substituent (Modified Shoolery Rules)

Note:

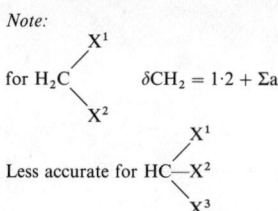

for $H_2C\overset{X^1}{\underset{X^2}{\diagdown}}$ $\delta CH_2 = 1.2 + \Sigma a$

Less accurate for $HC\overset{X^1}{\underset{X^3}{-X^2}}$

X	a	X	a
—=	0·75	—Ph	1·3
—≡	0·9	—Br	1·9
—COOH, —COOR	0·7	—Cl	2·0
—CN, —COR	1·2	—OR, —OH	1·7
—SR	1·0	—OCOR	2·7
—NH₂, —NR₂	1·0	—OPh	2·3
—I	1·4		

TABLE 5.3 δ Values for H Attached to Unsaturated and Aromatic Groups

$H\!-\!\!\equiv\!-R$	1·8*		5·6
$H\!-\!\!\equiv\!\!-\!\!OH$	2.4*		5·8
$H\!-\!\!\equiv\!-\!=\!-$, etc.	2,7*		
$H\!-\!\!\equiv\!-Ph$	2.9*		
$H\!-\!\!\equiv\!-CO\!-$	3·2*		6·0
$H_2C\!=\!C\overset{R}{\underset{R'}{\diagdown}}$	4·6		
$H_2C\!=\!-\!=\!-$, etc.	4·9		6·2

TABLE 5.3—*continued*

Structure	δ	Structure	δ
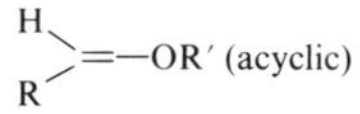 $\underset{R}{\overset{H}{>}}=$—OR′ (acyclic)	5·0	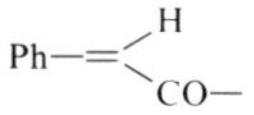 Ph—$=\overset{H}{\underset{CO-}{<}}$ (*cis* or *trans*)	6·6
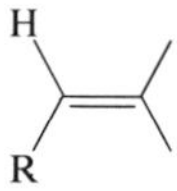 $\underset{R}{\overset{H}{>}}=<$	5·3	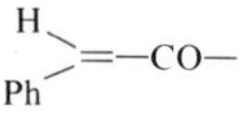 $\underset{Ph}{\overset{H}{>}}=$—CO—	7·8
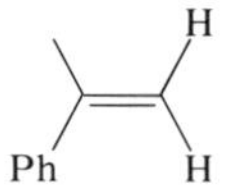 $>=<\overset{H}{}$ Ph H	5·0 / 5·3	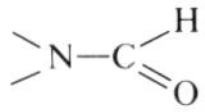 $>N-C\overset{H}{\underset{O}{<}}$	7·8
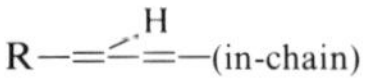 R—=—=—(in-chain) with H	6·2	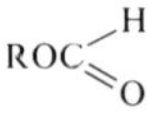 $ROC\overset{H}{\underset{O}{<}}$	8·0
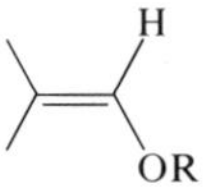 $>=<\underset{OR}{\overset{H}{}}$	6·8	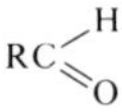 $RC\overset{H}{\underset{O}{<}}$	9·6
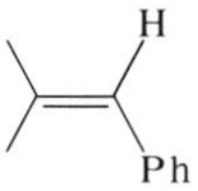 $>=<\underset{Ph}{\overset{H}{}}$	7·0	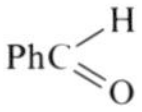 $PhC\overset{H}{\underset{O}{<}}$	9·9
Ph—H	7·27 (see table 5.5)		

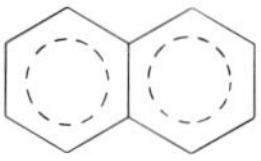

α 7·7 β 7·5

γ 7·4 β 7·0 α 8·5 (N)

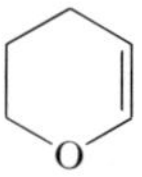

4·5 6·2 (O)

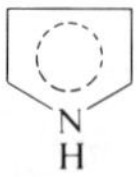

β 6·1 α 6·5 (N-H)

β 7·1 α 7·2 (S)

β 6·3 α 7·4 (O)

* Alkyne proton signals removed on deuteration, and δ values increased by a trace of pyridine.

TABLE 5.4 Proton–proton Spin-coupling Constants

Function	J_{ab}/Hz
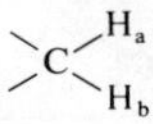 $\diagdown C \diagup \overset{H_a}{\underset{H_b}{\diagdown}}$ (gem)	10–18 depending on the electronegativities of the attached groups
$\diagdown CH_a{-}CH_b \diagdown$ (vic)	depends on dihedral angle: at $0°\ J = 8$ Hz; at $90°\ J = 0$ Hz; at $180°\ J = 9$ Hz.
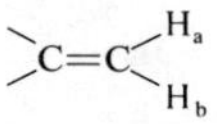 $\diagdown C{=}C \diagup \overset{H_a}{\underset{H_b}{\diagdown}}$	3–7
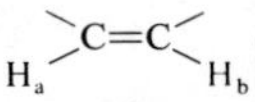 $\underset{H_a}{\diagdown} C{=}C \underset{H_b}{\diagup}$ (cis)	5–14
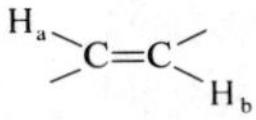 $\overset{H_a}{\diagdown} C{=}C \diagdown H_b$ (trans)	11–19
$\diagdown C{=}C \diagdown \overset{C{-}H_a}{\underset{H_b}{\diagdown}}$	4–10
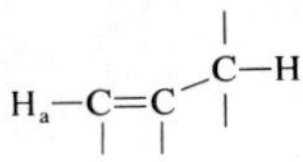 $H_a{-}C{=}C \diagdown C{-}H_b$ (cis or trans)	0–2 (for aromatic systems, 0–1)
$\diagdown C{=}CH_a{-}CH_b{=}C \diagdown$	10–13
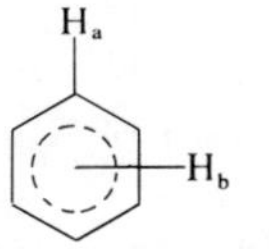 $\overset{H_a}{\bigcirc}{-}H_b$	ortho, 7–10 meta, 2–3 para, 0–1

TABLE 5.5   Shifts in the Position of Benzene Protons $(7 \cdot 27 \, \delta)$ caused by Substituents

Substituent	ortho	meta	para
—CH$_3$	−0·15	−0·1	−0·1
—=	+0·2	+0·2	+0·2
—COOH, —COOR	+0·8	+0·15	+0·2
—CN	+0·3	+0·3	+0·3
—CONH$_2$	+0·5	+0·2	+0·2
—COR	+0·6	+0·3	+0·3
—SR	+0·1	−0·1	−0·2
—NH$_2$	−0·8	−0·15	−0·4
—N(CH$_3$)$_2$	−0·5	−0·2	−0·5
—I	+0·3	−0·2	−0·1
—CHO	+0·7	+0·2	+0·4
—Br	0	0	0
—NHCOR	+0·4	−0·2	−0·3
—Cl	0	0	0
—$\overset{+}{N}$H$_3$	+0·4	+0·2	+0·2
—OR	−0·2	−0·2	−0·2
—OH	−0·4	−0·4	−0·4
—OCOR	+0·2	−0·1	−0·2
—NO$_2$	+1·0	+0·3	+0·4
—SO$_3$H, —SO$_2$Cl, —SO$_2$NH$_2$, etc.	+0·4	+0·1	+0·1

6

The Electronic (Ultraviolet–Visible) Absorption Spectrum

Absorption of ultraviolet or visible light by an organic molecule involves changes in the electronic energy levels within the molecule. The most accessible regions in the spectrum (ultraviolet from 190 to 400 nm, visible from 400 to 800 nm) are associated with changes in the electronic structure of multiple bonds such as C=C, C=O, N=N, aromatic systems, etc.; organic chemists use ultraviolet and visible absorption spectroscopy principally for studying these functions, and each is discussed below.

The term 'electronic absorption spectrum' will be used throughout as a general term covering spectra recorded in the ultraviolet and/or visible region, from 190 to 800 nm.

No details about the theory of electronic absorption spectra are given, and it is assumed that the student has had an introduction to the subject. The short notes given summarize the essential facts which are necessary for recording and analysing the spectra, and should serve to remind the student of the principal features with which he should be familiar.

Suitable complementary texts are the books by Williams and Fleming, by Dyer, by Schwarz, and by Kemp, listed on page 39.

More extensive coverage of electronic absorption spectroscopy alone is given in:

A. E. Gillam and E. S. Stern, *An Introduction to Electronic Absorption Spectroscopy in Organic Chemistry*, Edward Arnold, London, 2nd Ed., 1957.

R. A. Friedel and M. Orchin, *Ultraviolet Spectra of Aromatic Compounds*, John Wiley, New York, 1951.

E. A. Braude, 'Ultraviolet Absorption and Structure of Organic Compounds', *Annual Reports of the Chemical Society*, **42**, 105, 1945.

A. I. Scott, *Interpretation of the Ultraviolet Spectra of Natural Products*, Pergamon Press, Oxford, 1964.

A large number of spectra catalogues is available: it is most useful for the present purpose to have access to those spectra which have a large amount of fine structure and which are therefore difficult to describe in tabular form. In particular, aromatic hydrocarbons and heterocyclic aromatic compounds may have extensive series of peaks (see fig. 6.1). Many hydrocarbon spectra are given in:

E. Clar, *Polycyclic Hydrocarbons*, Vols. 1 and 2, Springer-Verlag and Academic Press, Berlin and London, 1965.

Other aromatic spectra are reproduced in Friedel and Orchin, listed above. Selected spectra are published in two other catalogue series:

Sadtler Ultraviolet Spectra, Sadtler and Son Inc., Pennsylvania, and Heyden and Son, London.

D.M.S. UV Atlas of Organic Compounds, Verlag Chemie, Weinheim, and Butterworths, London, 1965 (Vol. 4 published 1968).

6.1   The Absorption Laws

Lambert's law states that the absorbance of a sample is proportional to the length of the optical path. *Beer's law* relates the absorbance to the concentration of the substance. The most useful way of combining these two laws (into what is sometimes called the Beer–Lambert law) gives:

$$\log \frac{I_0}{I} = \varepsilon c l \quad \text{or} \quad \varepsilon = \frac{\log (I_0/I)}{cl}$$

I_0 is the intensity of the incident light (or, for most organic work, the light intensity passing through the reference cell).

I is the light intensity transmitted through the sample solution.

$\log (I_0/I)$ is the *Optical Density*, D (also called the *Absorbance*, A). This is measured directly by modern double-beam spectrometers.

c is the concentration of solute (in moles/litre).

l is the path length, in cm.

ε is the *Molecular Extinction Coefficient* (also called the *Molar Absorptivity*), a constant for a particular compound at a given wavelength. (But see 6.5.)

If the molecular weight is not known, then ε cannot be calculated from the above equation, since the molar concentration is not known; the following alternative expression is used, where c is the concentration in g/100 ml.

$$E_{1\,cm}^{1\%} = \frac{\log (I_0/I)}{cl} \quad \text{thus} \quad \varepsilon = E_{1\,cm}^{1\%} \times \frac{M.Wt}{10}$$

An absorption spectrum is a plot of ε, $\log \varepsilon$, or $E_{1\,cm}^{1\%}$ (on the ordinate) against wavelength (on the abscissa). Frequency units may also be plotted on the abscissa, but these are seldom used in organic chemistry in electronic absorption spectroscopy.

6.2 Solvents

Most organic compounds will have their spectra recorded in ethanol or in hexane (or cyclohexane), but choice of solvent is often dictated by solubility considerations. Another important factor is the low wavelength cut-off for each solvent, below which the solvent itself absorbs appreciably. The commonest solvents (with the lower wavelength limit below which they should not be used) are: 95 per cent ethanol (205 nm); methanol (205 nm); hexane (200 nm); cyclohexane (195 nm); ether or dioxan (215 nm); chloroform (240 nm); water (195 nm).

All solvents must be of spectroscopic quality; in particular, commercial grade cyclohexane contains some benzene (with an absorption peak at 255 nm), as does commercial absolute alcohol, from which the water has been removed by azeotropic distillation with benzene.

6.3 Solvent Effects

Change of solvent may shift both the position and the intensity of a particular absorption band, and it is most important to quote the solvent used in all absorption spectroscopy work.

The wavelength of an absorption band recorded in hexane or cyclohexane may be 10–15 nm lower than in alcohol: this effect is minimal for conjugated dienes (non-polar) but is important in the case of α,β-unsaturated ketones, etc. (polar). The values given on the tables for λ_{max}, unless otherwise stated, refer to alcohol solutions: values of λ_{max} recorded in less polar solvents should have the following quantities added to make comparisons possible:

hexane or cyclohexane, 10–15 nm; ether or dioxan, 6 nm; chloroform, 1 nm.

Values of λ_{max} recorded in water should have 8 nm *deducted*.

Fine structure shown in non-polar solvents may be masked in polar solvents. This is especially true in aromatic compounds, where spectra recorded in hexane solution show more (and sharper) peaks than those recorded in alcohol.

6.4 Cells

For accurate work synthetic silica cells or matched natural silica cells should be used. Glass cells absorb strongly *ca.* 300 nm and are not very useful for organic work.

Never handle the optical surfaces of the cells.

Wipe off spillage from the cell surfaces using soft tissues only.

Clean the cells immediately after use by thorough rinsing with solvent.

Drain the cells dry on clean tissue or on a rack, or store under water. Do not rest the optical surfaces against any support.

6.5   Solution Preparation

Always prepare standard solutions, *ca.* 0·1 per cent, so that ε or $E_{1\,cm}^{1\,\%}$ can be calculated from the absorbance [log (I_0/I)] recorded by the spectrometer. Prepare the solution in a standard volumetric flask (e.g., 20 cm^3), but never heat a standard flask.

The extinction coefficients of weak bands will be measurable on this first solution, but the concentration may be too high to permit the measurement of strong absorption bands. Dilution is necessary, and the most convenient method is to dilute by a factor of 10: pipette 2 cm^3 of the original solution into a 20 cm^3 standard flask and make up to the mark; the concentration of this new solution is one-tenth of the original concentration.

Some compounds do not obey Beer's law, and while these deviations may be a minor effect, in other cases gross changes may appear in the spectrum on dilution. These extensive changes are most frequently brought about by ionization or dissociation processes, and only a relatively small proportion of organic molecules exhibit them (e.g., barbiturates).

6.6   Recording the Spectra

If you have access to a recording spectrometer, the trace obtained will usually be a plot of absorbance against wavelength. Non-recording spectrometers read out absorbance on a dial, to be plotted manually against wavelength.

In either case, calculate ε or $E_{1\,cm}^{1\,\%}$ for the maxima. Where numerous maxima are shown, as in aromatic compounds, it is sufficient to calculate ε for two or three maxima; the other maxima can have their ε values read off from the spectrum directly if ε is plotted on a linear scale (since the shape of the spectrum is the same as the plot of absorbance against wavelength obtained from the machine).

For many aromatic systems, the values of ε vary by factors of several thousands: a linear plot of ε is inconvenient, and therefore spectra of these compounds are virtually always a plot of log ε against wavelength. Logarithmic paper can be obtained commercially for this purpose (e.g., five log cycles on the ordinate, with a linear cm scale on the abscissa).

Spectra recorded as log ε against wavelength have quite different appearance from those plotted in the linear form. The former are foreshortened on the ordinate compared to the linear spectra, but the wavelength scale is unchanged; see fig. 6.1. Care must be taken to extract extinction coefficient data carefully from the spectrum.

Where full spectra cannot be reproduced for reference purposes, the extinction coefficient and wavelength position of each maximum on the spectrum may be listed in tabular form. The data on individual compounds which follow these notes are in this form, giving λ_{max}, ε_{max} (EtOH) or λ_{max}, log ε_{max} (hexane), etc.

6.7   Alkenes, Dienes, and Trienes

Although normal ultraviolet spectrometers (in contrast with vacuum ultraviolet spectrometers) are sensitive down to 190 nm, organic solvents effectively cut off the region below 200 nm; as a direct consequence of this cut-off, isolated C=C bonds, absorbing around 190 nm, are not easily studied.

Where two C=C bonds are in conjugation (—CH=CH—CH=CH—), i.e., in 1,3-dienes, absorption moves to longer wavelength. Thus butadiene in hexane solution absorbs at 217 nm, ε 20,900.

In the same way, a conjugated triene absorbs around 250 nm.

This is an extremely important observation, since the position of the absorption allows us to say with confidence whether two double bonds in a molecule are conjugated or separated. If a molecule contains three double bonds, the spectrum will distinguish compounds in which the three double bonds are isolated, from those in which two or three are in conjugation.

The effect of substituents on λ_{max} can be predicted with reasonable confidence by the empirical data listed in table 6.1. Taking butadiene as the parent, as each hydrogen atom is replaced by one of the substituents listed, λ_{max} moves to longer wavelength. (This set of rules is usually referred to as the Woodward rules: R. B. Woodward first enunciated them, although modifications and extensions have been implemented from time to time.)

Cyclohexa-1,3-diene is the parent for cyclic conjugated dienes with both double bonds in the same ring (homoannular): cyclic dienes in which the double bonds are in different rings (heteroannular) are more closely related in their absorption characteristics to the acyclic dienes.

6.8   Carbonyl Compounds

The presence of a C=O group in an organic molecule is unambiguously detected from the infrared spectrum.

Isolated C=O groups show only weak absorption from the $n \rightarrow \pi^*$ transition around 300 nm; these transitions have very low ε values (10–100) and are little used for diagnostic purposes when infrared evidence is available.

α, β-Unsaturated ketones and aldehydes show strong absorption in the electronic absorption spectrum due to the $\pi \rightarrow \pi^*$ transitions of the conjugated π electron system; the position of this absorption can be calculated for simple derivatives using the modified Woodward rules shown on table 6.2. Other bands of lower intensity may be present in the spectrum around 350 nm ($n \rightarrow \pi^*$ transitions).

The position of the C=O *str* absorption in the infrared spectrum of carbonyl compounds varies from conjugated to non-conjugated members, and the two classes can usually be distinguished by this variation (see infrared correlation chart 2). The differences in their ultraviolet absorption spectra are greater, however, and this method makes differentiation more certain.

2,4-Dinitrophenylhydrazones of ketones and aldehydes absorb at much longer wavelength than the parent carbonyl compound. For saturated aldehyde 2,4-D.N.P., λ_{max} is 356–360 nm; for saturated ketone 2,4-D.N.P., λ_{max} is 362–366 nm; for α,β-unsaturated ketones and aldehydes, calculate λ_{max} for the parent compound and add *ca.* 150 nm for the 2,4-D.N.P.

Aryl ketones and aldehydes cannot be detected with any degree of certainty from their electronic absorption spectra, other than by comparison with the spectra of known compounds. This point is restated later in the discussion on aromatic compounds (6.9).

The distinctions among ArCOR, ArCOAr, Ar(CH$_2$)$_n$COR, etc., is usually more reliably carried out from infrared evidence (see infrared charts 2) but satisfactory distinctions can be made by reference to the electronic spectra of model compounds shown in the spectra catalogues.

α,β-*Unsaturated acids*, *esters*, etc., are less satisfactorily studied than ketones and aldehydes, since many show maxima near 200 nm; accurate recording of λ_{max} and ε_{max} at these low values is not easy, but wherever possible an attempt should be made to compare the observed values with those calculated from table 6.2.

6.9   Benzene Derivatives

Benzene itself shows two series of absorption bands, around 200 nm (ε 8000) and around 255 nm (ε 230). The degree of fine structure shown in hexane solution diminishes markedly when the spectrum is recorded in ethanol; fig. 6.1 shows the long wavelength absorption band for benzene (which solvent was used?).

Substitution derivatives of benzene all show strong absorption in the ultraviolet, so that it is possible to detect benzenoid structures by this method; but the identification of substituent functional groups from the electronic absorption spectrum is so much more ambiguous than infrared evidence that it is of little practical value in this context.

Spectral data for a large number of benzene derivatives are given in

Friedel and Orchin (see page 139), and the identity of a simple benzene derivative can be confirmed by comparison with known spectra.

A few generalizations can be made about the way the spectrum changes when a substituent is added to the benzene nucleus.

Extension of the conjugation moves λ_{max} to longer wavelength, increases ε_{max}, and reduces the fine structure. This effect applies also to the introduction of groups capable of donating non-bonding pairs to the ring (OH, OMe, NH_2, etc.).

Maximum shifts occur when an electron-donating group is *para* to an electron-withdrawing group; thus *p*-nitrophenol absorbs at longer wavelength and higher ε than do *o*-nitrophenol or *m*-nitrophenol.

Phenate ions absorb at longer wavelength than free phenols (greater conjugation of oxygen lone pairs), while amine salts absorb at shorter wavelength than free amines (no contribution from the nitrogen lone pair).

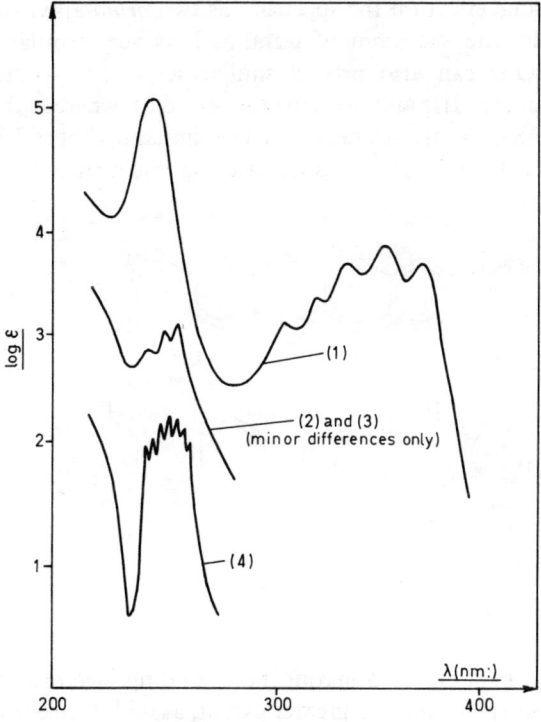

Fig. 6.1   Ultraviolet absorption spectra of (1) anthracene (2) 9,10-dihydroanthracene (3) anthracene/maleic anhydride adduct (4) benzene—Reproduced from W. Kemp, *Practical Organic Chemistry*, McGraw-Hill Publishing Company Ltd, London, 1967

6.10 Fused Aromatic Systems

The electronic spectrum is of great value in identifying the ring system in polynuclear aromatic compounds and in heterocyclic aromatic systems; the position of the absorption moves to long wavelength when the conjugation is extended by fusion of additional aromatic rings.

Table 6.3 lists a number of common hydrocarbons of this type; they are arranged according to the wavelength of the first (longest wavelength) absorption band so that an unknown hydrocarbon can readily be identified from the table.

Alkyl and halogen substituents on the ring alter the λ_{max} positions slightly, but the changes are so small that the ring system can be identified unequivocally from the electronic absorption spectrum. Thus 1-bromonaphthalene and 2-bromonaphthalene can be identified as naphthalene derivatives because the absorption spectra are very similar to that of naphthalene itself.

Note that the fusion of a *saturated* ring to an aromatic system produces roughly the same effect on the spectrum as two *ortho*-alkyl substituents; as an example, the spectrum of tetralin, I, is very similar to that of *o*-xylene, II. One can also predict similarities in the spectra of 9,10-dihydroanthracene, III, and *o*-xylene (see fig. 6.1), whereas the spectrum of 1,2,3,4-tetrahydroanthracene, IV, will be similar to that of 2,3-dimethylnaphthalene. In fig. 6.1, why are spectra 2 and 3 so similar?

Substituents capable of donating non-bonding electrons to the ring system alter the spectrum to a greater extent, as with benzene derivatives, and the position of substitution may affect the appearance of the spectrum considerably. This can be used, for example, to distinguish the α-series of naphthalene derivatives from the β-series, but peak-by-peak comparison with the spectra of known compounds is advisable.

6.11   Heterocyclic Aromatic Systems

Table 6.4 lists the absorption maxima of some common heterocyclic aromatic compounds. The effect of substituents on these spectra is similar to the effect observed in other aromatic systems (6.9 and 6.10). The number of possible cases is very large; the examples given are those most likely to be encountered in a course of qualitative organic analysis.

The electronic absorption spectrum of all heterocyclic compounds (and those suspected to be heterocyclic) should be recorded, and an attempt made to confirm the nature of the ring system from table 6.4.

Reference to spectra catalogues may be necessary in cases of doubt.

6.12   Azobenzenes

Alkyl-, alkoxy-, and halogen-substituted *trans*-azobenzenes have similar spectra to the parent azobenzene. They absorb well into the visible, showing three broad bands with approximate λ_{max} (log ε_{max}) as follows: 230 (4·2), 325 (4·3), 450 (2·6).

The *cis*-isomers, which are less likely to be encountered as student unknowns, in general show similar absorptions, often with lower intensities.

Substituents capable of conjugation or of lone pair donation shift the absorption to longer wavelength and higher intensity; this is the basis of azo dyes, whose structures (and thus colours) are altered by the introduction of a variety of substituents too numerous to be catalogued here.

The first absorption band of these compounds commonly appears at λ_{max} 500 nm, with log ε 3–4. Acidic and basic groups in the molecule will make the spectrum very dependent on pH; indicators such as methyl orange or methyl red are best identified by comparison of their spectra (at various pH values) with authentic samples.

TABLE 6.1   Conjugated Dienes and Trienes (in ethanol)
$\pi \rightarrow \pi^*$ Transitions, ε 10,000–20,000

Acyclic and heteroannular dienes	215 nm
Homocyclic dienes	253
Acyclic trienes	245
Addition for each substituent:	
—R alkyl (including part of a carbocyclic ring)	5 nm
—OR alkoxy	6
—Cl, —Br	25
—O.CO.R	0

Solvent shifts minimal

If one double bond is exocyclic to a ring, add 5 nm; if exocyclic to two rings, add 10 nm.

TABLE 6.2 α, β-Unsaturated Carbonyl Compounds (in ethanol)
$\pi \rightarrow \pi^*$ Transitions, ε 4500–20,000

Ketones $-\overset{\mid}{C}=\underset{\underset{\alpha}{\beta}}{\overset{\mid}{C}}-CO-$ Acyclic or 6-ring cyclic		215 nm
5-ring cyclic		202
Aldehydes $-\overset{\mid}{C}=\overset{\mid}{C}-CHO$		207
Acids and Esters $-\overset{\mid}{C}=\overset{\mid}{C}-CO_2H(R)$		197

Addition for each substituent:	α	β
—R alkyl (including part of a carbocyclic ring)	10 nm	12 nm
—OR alkoxy	35	30
—Cl	15	12
—Br	25	30
—O.CO.R	6	6

$-\overset{\mid}{C}=\overset{\mid}{C}-$ extension of the conjugation	30
Solvent shifts: see text, 6.3	
Exocyclic double bond: see Table 6.1	

TABLE 6.3 Aromatic Hydrocarbons

Values quoted are for λ_{max} in nm, with log ε_{max} in parentheses. Solvents used were either hexane (H) or ethanol (E).

Hydrocarbon	Solvent	Principal maxima
benzene	E	229 (1·21), 234 (1·46), 239 (1·76), 243 (2·00), 249 (2·30), 254 (2·36), 260 (2·30), 268 (1·04)
toluene, xylenes, and trimethylbenzenes		similar appearance to benzene spectrum; but peaks move to longer λ and higher ε with each additional alkyl group
biphenyl	E	250 (4·15); 2- or 2′-substituents may change the spectrum considerably
binaphthyls	H	220 (5·00), 280 (4·20)*; see biphenyl
indene	H	209 (4·34), 221 (4·03), 249 (3·99), 280 (2·69), 286 (2·35); many inflections
styrene	H	247 (4·18), 273 (2·88), 282 (2·87), 291 (2·76)
fluorene	E	261 (4·23), 289 (3·75), 301 (3·99)
naphthalene	E	221 (5·00), 248 (3·40), 266 (3·75), 275 (3·82), 285 (3·66), 297 (2·66), 311 (2·48), 319 (1·36)
acenaphthene	E	similar appearance to naphthalene spectrum
trans-stilbene	E	230 (4·20), 299 (4·48), 312 (4·47)
cis-stilbene	E	225 (4·34), 282 (4·10)
phenanthrene	E	223 (4·25), 242 (4·68), 251 (4·78), 274 (4·18), 281 (4·14), 293 (4·30), 309 (2·40), 314 (2·48), 323 (2·54), 330 (2·52), 337 (3·40), 345 (3·46)
phenalene (perinaphthene)	E	234 (4·40), 320 (3·90), 348 (3·70)
fluoranthene	E	236 (4·66), 276 (4·40), 287 (4·66), 309 (3·56), 323 (3·76), 342 (3·90), 359 (3·95)
chrysene	E	220 (4·56), 259 (5·00), 267 (5·20), 283 (4·14), 295 (4·13), 306 (4·19), 319 (4·19), 344 (2·88), 351 (2·62), 360 (3·00)
pyrene	E	231 (4·62), 241 (4·90), 251 (4·04), 262 (4·40), 272 (4·67), 292 (3·62), 305 (4·06), 318 (4·47), 334 (4·71), 352 (2·82), 362 (2·60), 372 (2·40)
anthracene	E	252 (5·29), 308 (3·15), 323 (3·47), 338 (3·75), 355 (3·86), 375 (3·87)
perylene	E	245 (4·44), 251 (4·70), 387 (4·08), 406 (4·42), 434 (4·56)
acenaphthylene	H	229 (4·72), 264 (3·46), 274 (3·43), 311 (3·93), 323 (4·03), 334 (3·70), 340 (3·70), 440 (2·00), 468 (1·56)

* 2,2′-binaphthyls are different: 255 (4·9), 320 (4·4).

TABLE 6.4   Heterocyclic Systems

Values quoted are for λ_{max} in nm, with log ε_{max} in parentheses. Solvents used were usually hexane (H) or ethanol (E); change of solvent may affect the spectrum considerably.

Compound	Solvent	Principal maxima
pyrrole	E	235 (2·7, shoulder); no sharp maxima
furan	H	207 (3·96)
thiophen	H	227 (3·83), 231 (3·85), 237 (3·82), 243 (3·58)
indole	H	220 (4·42), 262 (3·80), 280 (3·75), 288 (3·61)
carbazole	E	234 (4·63), 244 (4·38), 257 (4·29), 293 (4·24), 324 (3·55), 337 (3·50)
pyridine	H	251 (3·30), 256 (3·28), 264 (3·17)
quinoline	E	226 (4·53), 230 (4·47), 281 (3·56), 301 (3·52), 308 (3·59)
isoquinoline	H	216 (4·91), 266 (3·62), 306 (3·35), 318 (3·56)
acridine	E	249 (5·22), 351 (4·00)
pyridazine	H	241 (3·02), 246 (3·15), 251 (3·15), 340 (2·56)
pyrimidine	H	242 (3·31), 293 (2·51), 307 (2·40), 313 (2·18), 317 (2·04), 324 (1·73)
pyrazine	H	254 (3·73), 260 (3·78), 267 (3·57), 315 (2·93), 322 (2·99), 328 (3·02)
barbituric acids	water	256–7 (ca. 4.4, concentration-dependent)

7
The Mass Spectrum

Since mass spectrometry is now being applied routinely to the investigation of organic structures, it seems inevitable that, within a short time, all student courses in qualitative organic analysis must incorporate this technique. It is unlikely within this period that many students will have free access to expensive spectrometers, but mass spectrometric data can be obtained from published sources and supplied to the student as supplementary information; the interpretation of these data as a part of the analysis scheme is a valuable practice ground for later, more advanced, mass spectrometry studies.

The discussion which follows is obviously far from a comprehensive survey of all the information which can be derived from a mass spectrum, but rather comprises a set of introductory rules. The unavoidable simplification does not in any way infringe on the rules which must be followed in more exhaustive examinations of the mass spectrum.

Complementary texts dealing with the theory and practice of mass spectrometry are: the books by Williams and Fleming, by Schwarz, and by Kemp, referred to on page 39. More detailed discussions are given in a number of texts, for example:

F. W. McLafferty, *Interpretation of Mass Spectra*, W. A. Benjamin Inc., New York, 1966.

H. C. Hill, *Introduction to Mass Spectrometry*, Heyden and Son, London, 1966.

K. Biemann, *Mass Spectrometry. Organic Chemical Applications*, McGraw-Hill Book Co. Inc., New York, 1962.

A classified catalogue of mass spectral data is a valuable adjunct, for example: A. Cornu and R. Massot, *Compilation of Mass Spectral Data*, Heyden and Son, London, and Presses Universitaires de France, 1966. First Supplement, 1967, Second Supplement, 1971.

Further assistance in the interpretation of molecular formulae from molecular weight etc., can be obtained from: J. Lederberg, *Computation of Molecular Formulas for Mass Spectrometry*, Holden-Day, San Francisco, 1964.

You are likely to be presented with mass spectrometric data in one of three ways. The most comprehensive presentation (but not the clearest) is obtained directly from the machine, and this may be three to six simultaneous traces of the m/e ratios recorded at differing sensitivities. This should be recoded into the most useful presentation, the line diagram (see fig. 7.1) in which the m/e ratios are plotted against relative abundance. To do this, examine the trace from the machine which has been recorded at lowest sensitivity, and identify the most abundant peak; this is named the *base peak* (the 105 peak in fig. 7.1), and the intensities of all the other peaks are measured as a percentage of the base peak intensity. Plot these percentages against m/e ratios.

The third method of presenting mass spectrometric data is simply a list of the most abundant peaks, together with their relative abundances; usually the ten or twenty most abundant peaks are specified, so that this is the least comprehensive presentation.

Whatever the presentation, the following process is intended to identify from the mass spectrum certain characteristic groupings; in particular you will be looking for fragmentations which support your interpretation of the presence of alkane groupings, aromatic residues, and functional groups. Where possible the molecular formula will also be deduced.

7.1   The Molecular Ion

Not all mass spectra show the ion corresponding to loss of an electron from the complete molecule, $M^{\ddot{+}}$; compounds whose mass spectra do not show the molecular ion (or parent peak) may have highly branched structures. Abundant molecular ions are given by most aromatic compounds (and many heterocyclics).

Identification of the molecular ion should be carried out carefully, and its identity must be confirmed by applying a number of tests:

(a) First note the peak of highest m/e ratio; in most cases this will also be the molecular ion. An exception to this concerns compounds which give a particularly stable, and hence abundant parent peak. If the $M^{\ddot{+}}$ peak is intense, then it is possible to observe $M + 1$ peaks; in a few cases, particularly with Cl or Br present, $M + 2$ may also be seen.

$M + 1$ peaks arise principally from the natural abundance (1·1 per cent) of ^{13}C in organic compounds: an organic molecule containing one ^{13}C atom will have a molecular weight one unit higher than the molecule containing only ^{12}C atoms. The presence of ^{2}H or ^{17}O has the same effect but statistically this is a less likely occurrence (see later). Compounds containing n carbon atoms will show an $M + 1$ peak whose intensity is about $1·1n$ per cent of the $M^{\ddot{+}}$ intensity: a more exact calculation is given later in section 2. Although n will not be known accurately until the compound has been identified, the value of M^+ enables you to deduce an approximate value for n which *in most instances* is sufficiently exact. The observed abundance of the $M + 1$ peak may be greater than predicted because of ion-molecule reactions, but it should not be substantially less than predicted.

The same considerations apply to abundant fragment ions: the principal isotope peak (p) is always associated with a $p + 1$ peak.

$M + 2$ peaks arise most commonly because of the presence of Cl or Br in the molecule, since the natural abundance of heavier isotopes in these elements is high (24·2 per cent for ^{37}Cl, and 49·48 per cent for ^{81}Br). Sulphur compounds may also show $M + 2$ peaks, but their intensity will be small unless more than one S atom is present. The natural abundance of ^{34}S is 4·39 per cent.

(b) If the molecule does not contain nitrogen, the molecular weight *must* be even: if an *even* number of nitrogen atoms is present, the molecular weight is also even. An *odd* molecular weight only arises with an *odd* number of nitrogen atoms in the molecule.

(c) Look for the following peaks below the apparent molecular ion: $M - 4$ or $M - 5$ (multiple loss of hydrogen is rare); $M - 14$ (loss of CH_2 is extremely rare). If these peaks are present below the ion of highest m/e value, then this ion is not likely to be M^+.

(d) Additional confirmation of the identity of M^+ will be given by the appearance of common fragment ions associated with particular functional groups as discussed below, for example $M - 15$ (loss of CH_3), $M - 29$ (loss of C_2H_5), etc.

7.2   Molecular Formula from Molecular Ion

If the molecular ion has been identified, then reference to published tables or a process of calculation will give all the possible molecular formulae which satisfy this molecular weight. This is particularly true where high-resolution data are available, and the molecular weight is known to five or six decimal places (see Lederberg).

If only low-resolution data are available, an additional check on the molecular formula can be carried out by again using the known relative abundances of the heavier isotopes in the organic molecule.

As an example, if the molecular weight is 162, then the molecular formula might be $C_9H_6O_3$ or $C_{11}H_{14}O$. In the former, with 9 carbon atoms, 6 hydrogen atoms, and 3 oxygen atoms, and with the natural abundances of ^{13}C, ^{2}H, and ^{17}O being respectively 1·1, 0·016, and 0·04 per cent, then the intensity of the $M + 1$ with respect to the M^+ peak will be $(9 \times 1·1 + 6 \times 0·016 + 3 \times 0·04) = 9·306$ per cent.

By a similar calculation, find the expected intensity of the $M + 1$ peak in the second compound.

It ought to be possible to differentiate these two, provided the molecular ion is relatively abundant, and provided no complication arises from ion-molecule reactions. A calculation of the expected intensity of the $M + 2$ peak can also be made, but this may be no more helpful in confirming the molecular formula than (e.g.) expected differences in the fragmentation patterns.

7.3   Fragmentation Patterns

The fragmentation modes of organic compounds are extremely complex, and it is not always easy to write structures for the ions produced in the mass spectrometer which are convincing in terms of 'ground state' chemistry. The fragmentation may be affected considerably by the functional groups present in the molecule—some functional groups affect the fragmentation more than others: the following notes indicate the way in which the commonest functional classes undergo fragmentation.

Functional groups are presented in the order met earlier in the book, beginning with compounds containing C, H, and O only.

Monofunctional compounds only are discussed; polyfunctional compounds will give more complex fragmentations which in many instances can be treated as a composite of the fragmentations for each individual functional class.

First pay some attention to the structure of the base peak, and of other outstandingly abundant ions. Fragmentation occurs easily at branching points in the molecule, so that, for example, if a peak at m/e 57 is very abundant (this may correspond to $C_4H_9{}^+$), the molecule may contain a butyl group attached to a highly substituted carbon atom.

Metastable Ions
Many aromatic compounds give rise to the phenyl cation (m/e 77), which loses $CH{\equiv}CH$ to give an ion of m/e 51. This fragmentation takes place during the acceleration of the m/e 77 ion along the spectrometer, and the instrument records the presence of an additional peak (m^*) whose m/e ratio is neither 77 nor 51, but is given by

$$m^* = \frac{51^2}{77} = 33 \cdot 8$$

In other words, because m/e 77 fragments during acceleration directly (i.e., in one step) to m/e 51, we observe also a *metastable peak*, as evidence of this fragmentation step. The presence of a metastable peak in the mass spectrum is a very valuable link between two of the fragment ions. (See below, 7·5 and 7·6.) Remember, however, that not *all* one-step fragmentations need show a metastable ion.

Metastable peaks are easily identified on the mass spectrum trace obtained directly from the spectrometer, as they are broad peaks rather than sharp lines. Where only numerical mass spectrum data are given, metastable peaks will be specifically indicated. In all cases, use the m/e value of the metastable peak to calculate which two fragment ions are inter-related: when ion a fragments to ion b the m/e value of the metastable ion (if one is formed) will be found at a value lower than that of b namely ($m^* = b^2/a$).

7.4 Aliphatic Compounds

If infrared (and n.m.r.) evidence indicate the presence of aliphatic groups, this can be confirmed in the mass spectrum; look for the common alkyl ions formed when aliphatic chains rupture:

m/e value	15	29	43	57	71 etc.
structure	CH_3^+	$C_2H_5^+$	$C_3H_7^+$	$C_4H_9^+$	$C_5H_{11}^+$

For n-alkanes, the R^+ ions reach maximum abundance around $C_3H_7^+$ and $C_4H_9^+$, and thereafter tail off quickly in abundance: for other alkanes, any peak which stands out above the normal distribution (i.e., a peak of abnormally high abundance) indicates a branching point in the chain.

With the exception of CH_3^+, these peaks are always associated with lesser amounts of the corresponding alkenyl ions, which contain two hydrogens less, at m/e 27, 41, 55, 69.

Look for peaks arising from the *loss* of these alkyl groups (as radicals) from the molecular ion: $M - 15$, $M - 29$, $M - 43$, etc. Note again the relative abundances of these ions, as this is related to the ease of fragmentation of the molecule.

7.5 Aromatic Compounds

The simplest aromatic compounds are those which contain a monosubstituted benzene ring, but we must also consider other carbocyclic aromatic systems (naphthalene etc.). Heterocyclic compounds are discussed later with compounds containing N, S, etc.

If the molecule contains also an aliphatic chain, then we shall find peaks in the mass spectrum as described above under aliphatic compounds.

The molecular formula for all aromatic compounds must have at least four double-bond equivalents (one for the ring and three for the 'unsaturation'). See 7.12.

Look for the peak at m/e 77, the phenyl cation; this is usually associated with m/e 51 (produced by elimination of CH≡CH from the phenyl cation) and by the metastable peak at m/e 33·8 (the position being calculated from $51^2/77$).

Look also for the peak at m/e 91, represented as I; if this is present as an intense peak, then the compound contains a benzene ring with an alkyl group attached. A peak at m/e 92 may be the methylenecyclohexadiene radical ion III; if this is present, it is strong evidence for a benzene derivative with a side chain of at least three carbon atoms (II), undergoing a β-elimination (see Ketones, 7.6).

I (tropylium)   II   III
m/e 91 m/e 92

The peak at m/e 92 may, however, be the $p + 1$ peak from the tropylium ion at m/e 91: in this case the 92 peak should be *ca.* 8 per cent ($7 \times 1\cdot1$) of the 91 peak intensity.

In the case of naphthalene derivatives and other polynuclear aromatic compounds, the molecular ion is often very stable and is frequently the base peak. The ability to delocalize charge over the molecule is so great in polynuclear hydrocarbons, that doubly charged ions arise occasionally in their mass spectra; this is discussed later under hydrocarbons.

7.6 Ketones

α-*Cleavage* of RCOR′ will give RCO$^+$ and R′CO$^+$, together with R$^+$ and R′$^+$. These may vary widely in relative intensities depending on stabilities. For example, in the spectrum of PhCOR, the two most intense peaks are usually PhCO$^+$ (m/e 105) and Ph$^+$ (m/e 77); loss of CH≡CH from Ph$^+$ gives the peak at m/e 51 mentioned in 7.5. Metastable peaks frequently appear with the sequence $105 \longrightarrow 77 \longrightarrow 51$; their presence at m/e 56·5 and m/e 33·8 is powerful confirmation of the PhCO group in the molecule. The occurrence of metastable peaks is indicated on fig. 7.1 by an asterisk.

β-*Cleavage* involves rupture of the α,β bond, together with H-abstraction from the γ-carbon atom of an aliphatic chain.

IV

The McLafferty rearrangement

This is an extremely common rearrangement process in carbonyl compounds, and frequently gives rise to the base peak; it is often referred to as the McLafferty rearrangement (see F. W. McLafferty's book, page 123). If R or R' is C_2H_5 or higher, IV may undergo a second β-elimination of an alkene, and produce peaks corresponding to IV $-$ 28 (loss of $CH_2{=}CH_2$), IV $-$ 42 (loss of $CH_3CH{=}CH_2$), etc.

Common values for IV itself are m/e 58 (R = CH_3, R' = H); 72 (R = R' = CH_3, or R = C_2H_5, R' = H); 120 (R = Ph, R' = H), etc. All of these fragments appear at even m/e values.

7.7 Aldehydes

α-*Cleavage* of RCHO can give rise to RCO^+, R^+, or HCO^+ (m/e 29), this last being weak from C_4 aldehydes upwards; note that $C_2H_5{}^+$ also appears at m/e 29, so that higher aldehydes showing a prominent m/e 29 peak do so because of $C_2H_5{}^+$ and not because of HCO^+. Loss of H$^.$ from the molecular ion is energetically unfavourable, so that abundant RCO^+ ($M - 1$) ions are only found where the charge can be effectively delocalized (R = Ph, etc.). Further comment, particularly about the PhCO group, is included in 7.6.

β-*Cleavage* can occur under the same circumstances as for ketones giving IV, where R must be H; m/e values are as for ketones, but one additional member of the series is possible at m/e 44 (R = R' = H). Peaks may also arise at $M - 44$, which represents charge retention by the eliminated alkene.

7.8 Quinones

As in ketones, α-cleavage occurs readily, and both α bonds can rupture as in V and VI:

V   VI

Peaks can arise in these cases at m/e 54 and at $M - 54$; anthraquinone and phenanthrenequinone show abundant ions due to loss of two molecules of CO ($M - $ CO at m/e 180, $M - $ 2CO at m/e 152); an abundant ion is also shown at m/e 76 ($C_6H_4{}^+$).

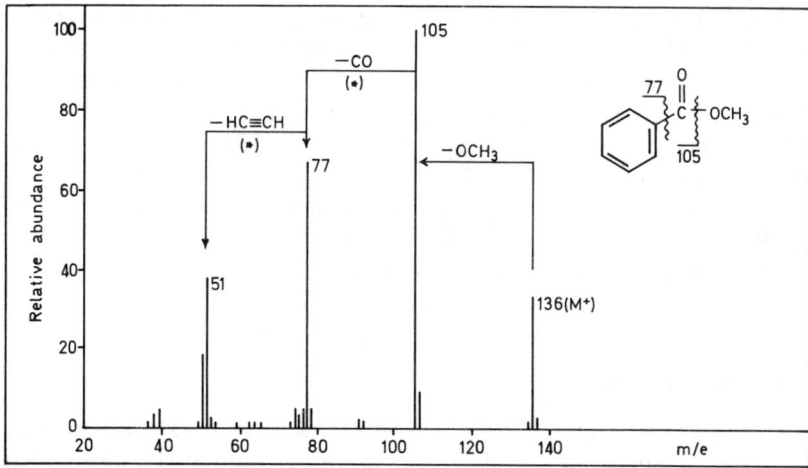

Fig. 7.1   Mass spectrum of methyl benzoate. Reproduced from D. H. Williams and I. Fleming, *Spectroscopic Methods in Organic Chemistry*, McGraw-Hill Publishing Company Ltd, London, 1966

7.9   Esters

Ester fragmentation is very similar to ketone behaviour; α- and β-cleavages give rise to peaks due to R^+, RCO^+, $ROCO^+$, and to variants of IV ($R=CH_3O$, C_2H_5O, PhO, etc.) (see fig. 7.1).

Common m/e values for R^+ and RCO^+ are given under ketones (7.6): for $ROCO^+$, peaks arise at m/e 59 (CH_3), 73 (C_2H_5), etc.

Common m/e values for the β-cleavage products of esters (IV) are 74 ($R = CH_3O$), 88 ($R = C_2H_5O$), etc. Where $R = C_2H_5O$, i.e., in ethyl esters, a second β-cleavage with loss of ethylene from IV gives rise to a peak at m/e 60.

Esters of aryl acids usually give strong $ArCO^+$ peaks (see under ketones, 7.6), and may also show $ArCO_2H^+$ peaks. Acetate esters may show loss of acetic acid ($M - 60$), but this is less important in phenolic acetates: the base peak in phenyl acetate itself is at m/e 94 ($PhOH^+$), which then loses CO to give m/e 66. This is discussed below under ethers, 7.11.

7.10   Alcohols

The mass spectra of all alcohols show very weak parent peaks; tertiary alcohols may show none. Fragmentation of the molecular ion can occur by elimination of alkyl or H radicals, the ease of elimination increasing with size and branching ($H^{\cdot} < CH_3^{\cdot} < C_2H_5^{\cdot}$, etc.). This means that the common peaks in the spectrum correspond to $M - C_3H_7$ ($M - 43$), $M - C_2H_5$ ($M - 29$), $M - CH_3$ ($M - 15$): $M - H$ is a small peak.

Long-chain alcohols may show peaks corresponding to $M - 1$, $M - 2$, and $M - 3$:

$$M-1 \qquad\qquad M-2 \qquad\qquad M-3$$

Loss of H_2O ($M - 18$) may also be accompanied by the simultaneous loss of an alkene, for example C_2H_4 (28), giving a peak at $M - 46$.

Alcohols with phenyl groups attached to C—OH give fragments in which the aryl group favours charge retention, rather than elimination as a radical; thus in 1-phenylethanol the base peak corresponds to $M - CH_3$, and peaks for $M - 43$, $PhCO^+$, and Ph^+ are also shown.

For these alcohols, a peak at m/e 107 (e.g., the $M - CH_3$ peak in the spectrum of 1-phenylethanol) is best represented as the hydroxytropylium ion VII, which loses CO to give VIII (m/e 79) and then loses H to give the phenyl cation. Further fragmentation of Ph^+ gives m/e 51 as pointed out in 7.6.

$$\text{VII} \qquad\qquad \text{VIII}$$
$$m/e \ 107 \qquad\qquad m/e \ 79$$

7.11   Ethers

Aliphatic ethers give weak $M^{\ddagger}$ peaks; this ion then loses alkyl radicals by breaking of the bond shown in IX, ease of radical elimination having been discussed in 7.10.

$$\text{IX} \qquad\qquad\qquad \text{X}$$

Peaks will therefore arise at m/e values corresponding to loss of CH_3˙ ($M - 15$), loss of C_2H_5˙ ($M - 29$), etc.

The C—O bond fission process indicated in XI occurs to a small extent only; where it occurs, charge retention by carbon is more favoured than by oxygen.

In ions such as Xa, however, C—O bond fission will be favoured if it is possible for a hydrogen atom to migrate from the β carbon atom on to oxygen: when this takes place, an alkene is eliminated and an oxonium ion XII produced:

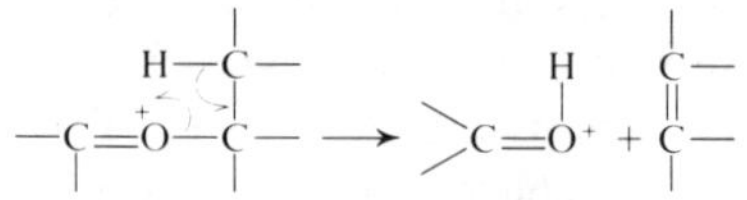

Common m/e values for XII are: 45 (CH_3 and H as substituents), 59 (C_2H_5, H), 73 (C_3H_7, H), etc. Note that all must have odd m/e values.

In aromatic ethers a similar process can be envisaged to explain the m/e 94 peak in the spectrum of many phenyl ethers: if β hydrogen is present, and this therefore excludes methyl ethers, then the molecular ion can eliminate an alkene as above. For the simplest case, phenetole, ethylene is eliminated leaving $PhOH^{·+}$ (XIII, m/e 94). This then loses CO to afford XIV, m/e 66.

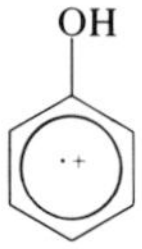

XIII	XIV	XV	XVI
m/e 94	m/e 66	m/e 93	m/e 65

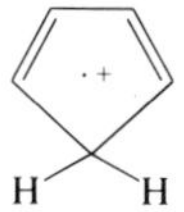

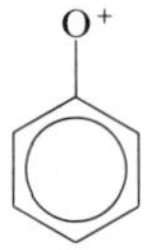

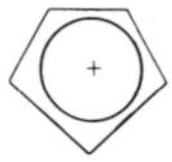

For methyl aryl ethers, the molecular ion can lose either HCHO $(M - 30)$, or $CH_3^·$ $(M - 15)$: loss of $CH_3^·$ produces a species such as XV, which then loses CO giving XVI.

Furan fragments to produce ions corresponding to $HCHO^+$ $(m/e$ 29), and also to loss of HCHO $(M - 29)$ and CO $(M - 28)$. Substituents in the furan ring will of course alter the importance of these fragmentations.

7.12 Hydrocarbons

Fragmentation of the carbon residues of hydrocarbons has been discussed in a general way in 7.4 and 7.5, and nothing further will be added on saturated hydrocarbons.

Use the molecular formula to infer unsaturation, ring formation, and aromaticity: this can be expressed as *double bond equivalents*, D.B.E.

For example, alkanes have formulae C_nH_{2n+2}; cycloalkanes and monoenes, C_nH_{2n}, 1 D.B.E.; bicycloalkanes, dienes, cycloalkenes, and alkynes, C_nH_{2n-2}, 2 D.B.E.; cyclodienes, trienes, and enynes, C_nH_{2n-4}, 3 D.B.E.; alkylbenzenes, C_nH_{2n-6}, 4 D.B.E.; naphthalenes, C_nH_{2n-12}, 7 D.B.E.; anthracenes, C_nH_{2n-18}, 10 D.B.E., etc.

Alkenes undergo many of the chain fragmentations of saturated hydrocarbons, but rupture of the allylic bond is favoured because of the stability of the allyl cation. Since the molecular ion ruptures so easily, parent peaks are often small.

$$\text{R—CH}_2\text{—CH=CHR} \longrightarrow \text{R}\overset{\displaystyle\frac{}{}}{}\text{CH}_2\text{—CH}\overset{+}{\underset{\cdot}{—}}\text{CHR} \longrightarrow \overset{+}{\text{CH}}_2\text{—CH=R} + \text{R}^{\cdot}$$
$$M^{+} \hspace{5cm} \text{XVII}$$

Common values for XVII are: m/e 41 (R = H), m/e 55 (CH_3), m/e 69 (C_2H_5), etc.

Cyclohexene derivatives have abundant $M^{\ddagger}$ ions which undergo the retro-Diels–Alder reaction, so that the molecular ion decays to an alkene and a 1,3-diene: charge retention depends on substituents, and either the alkene or the diene may carry the charge. Thus:

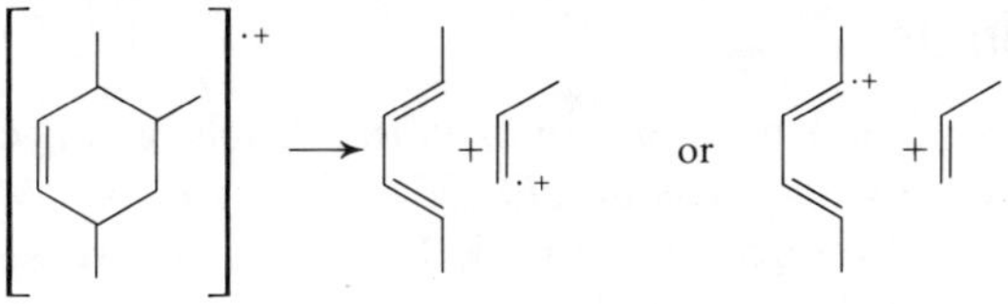

In aromatic hydrocarbons, it is most important to note the high relative abundance of the parent peak; $M^{\ddagger}$ is often the base peak, and $M + 1$ and $M + 2$ peaks are easily seen. Other peaks may be relatively small compared to the $M^{\ddagger}$ peak, but alkylbenzenes and alkylnaphthalenes rupture easily to produce abundant tropylium ion peaks (see 7.5).

Doubly charged ions may be produced. Since the molecular weight will be *even*, $M/2e$ will be an integer and it will not be possible to distinguish this peak from other simple fragment ions; but the isotope peak at $(M + 1)/2e$ will appear at a non-integer value and will be easily distinguished. The same arguments apply to doubly charged fragment ions.

7.13 Carboxylic Acids

Aliphatic acids capable of undergoing β-cleavage (see 7.6) will produce derivatives of IV, where R = OH; peaks at m/e 60, 74, etc., are common. A characteristic peak due to CO_2H^{+} appears at m/e 45, and $M - 45$ peaks may also arise.

Aryl acids lose OH, leaving the aroyl cation $ArCO^{+}$ (e.g., $PhCO^{+}$ at m/e 105) which loses CO then CH≡CH; this is discussed under ketones, in 7.6.

7.14   Carboxylic Acid Anhydrides

For saturated acyclic anhydrides, the base peak usually corresponds to RCO^+ (m/e 43, 57, 71, etc.) unless the chain is highly branched, in which case chain rupture may lead to an absence of the complete acyl group RCO^+ from the spectrum. Molecular ion peaks are weak or absent. Other fragmentations are variable in their occurrence, but the following are common: m/e 60 or $M - 60$ ($CH_3CO_2H = 60$); McLafferty rearrangement gives m/e 44, 58, etc.; m/e 42 ($CH_2{=}CO^+$).

Cyclic aliphatic anhydrides show a strong or base peak corresponding to loss of CO and CO_2 from M^+ ($M - 72$); other anhydride fragmentations are less characteristic, but analogous to the acyclic anhydrides.

Aromatic acyclic anhydrides fragment mainly to $ArCO^+$ etc., but $ArCO_2H^+$ and $M - CO$ are also seen. Cyclic aromatic anhydrides lose CO and CO_2 from M^+, and loss of H from $ArCO^+$ may also occur (e.g., m/e 104 instead of 105).

7.15   Phenols

Two alternative fragmentations are open to phenols, giving characteristic peaks. The first involves loss of CO ($M - 28$) or CHO ($M - 29$). The second fragmentation occurs if an alkyl group is also present, as in the cresols; such phenols undergo benzylic fission, leaving the hydroxytropylium ion VII at m/e 107. This fragments further as discussed under alcohols, 7.10.

7.16   Nitro Compounds

Several fundamental fragmentation sequences may be observed in aryl nitro compounds. Loss of NO_2 ($M - 46$) is general, followed by the normal fragmentations of the aryl cation. Successive loss of NO ($M - 30$) and CO ($M - 58$) can also occur. (Loss of NO leaves a PhO^+ ion, which indicates that the M^+ ion must isomerize in the spectrometer to PhONO.)

If an *ortho*-substituent contains hydrogen, then this may be eliminated with an oxygen atom from the nitro group ($M - OH$, i.e., $M - 17$), to be followed by the loss of CO ($M - 45$). This last ion may have the structure XVIII.

7.17   Amines

An odd number of nitrogen atoms in the molecule means an odd molecular weight.

Primary aliphatic amines rupture principally to give an alkyl radical and

$CH_2=NH_2{}^+$ (m/e 30) as the base peak: loss of smaller radicals from the alkyl chain gives the higher homologues of this ion at m/e 44, 58, 72, etc.) in decreasing amounts. Elimination of an alkene (e.g., $M - 28$) may also be observed.

With N-alkyl secondary and tertiary amines (including saturated heterocyclics), a similar fragmentation arises. The largest alkyl group leaves as a radical, and the ion produced has an m/e value corresponding to $M - R$. Further elimination of an alkene may occur, with hydrogen transfer to the nitrogen atom, and the m/e 30 ion commonly appears in these spectra (see above).

Primary aryl amines lose HCN to give $M - 27$ ions, whose structures can be represented by XIX. $M - 1$ peaks are also common, and many primary amines show m/e 106. N-Alkylanilines undergo α-cleavage of the alkyl group, the ion having m/e 106 (XX).

XVIII	XIX	XX
		m/e 106

Aromatic heterocyclic bases such as pyridine and quinoline give strong $M^{\ddagger}$ peaks. Alkyl substituted derivatives behave like alkylbenzenes and rupture at the 'benzylic' bond, the ion probably being an aza analogue of the tropylium ion I: m/e values correspond to $M - R$. Thereafter HCN splits off, and m/e values for these ions correspond to $M - R - 27$. In pyridine and quinoline themselves, this is the major fragmentation mode, i.e., loss of HCN, $M - 27$.

7.18 Amides

Aliphatic primary amides eliminate the alkyl group as R $\cdot$, leaving the ion $CONH_2{}^+$ at m/e 44. Wherever possible, β-cleavage also occurs (see 7.6) giving the ion IV, with R $= NH_2$, at m/e 59, 73, etc.

Secondary and tertiary amides behave similarly, giving $M - R$ peaks (e.g., $(CH_3)_2N{-}CO^+$) by α-cleavage. Subsequently, where β-cleavage is possible, various substitution derivatives of IV are produced with R $= NHCH_3$, $N(CH_3)_2$, etc.

Primary amides also show $M - NH_2$ peaks ($M - 16$).

Primary aryl amides fragment principally by loss of NH_2, to give the now familiar $ArCO^+$ ions: further fragmentation of this ion is discussed in 7.6.

7.19 Nitriles

Loss of HCN ($M - 27$) only appears in the lower aliphatic nitriles. From C_4 upward, the fragmentation is dominated by β-fragmentation analogous to that of ketones, etc. (7.6). Peaks corresponding to $CH_2{=}C{=}NH^+$ and its homologues therefore appear prominent at m/e 41, 55, 69, etc.

In aryl nitriles, loss of HCN may again occur, but nitriles of alkylbenzenes will rupture preferentially at the benzylic position; this may be followed by elimination of HCN.

7.20 Halogen Compounds

The most striking feature of the mass spectra of compounds containing Cl or Br is associated with isotope abundance. Since the abundance of ^{79}Br and ^{81}Br are almost the same (50·52 : 49·48), any ion containing one bromine atom will appear as a pair of peaks of equal intensity, with a separation of 2 mass units. *These doublets are easily noted on the spectrum.* For chlorine ($^{35}Cl : ^{37}Cl = 75·8 : 24·2$) doublets also arise, spaced at 2 mass units, the lower m/e peak being three times the intensity of the higher.

These arguments also apply to the molecular ion, which therefore appears as a doublet at $M^{\ddagger}$ and $M + 2$.

Aliphatic halides show fragmentation which is typical of the alkyl chain, but with additional features as follows.

Chlorine compounds: fragmentation by loss of HCl ($M - 36$ and $M - 38$) is more favoured than loss of Cl^+ (m/e 35 and 37). Peaks may also be seen corresponding to HCl^+ (m/e 36 and 38), and to loss of $Cl^·$ ($M - 35$ and $M - 37$).

Bromine compounds: fragmentation is similar to chlorides, but here loss of $Br^·$ is the preferred fragmentation.

Iodine compounds: these are similar to bromides, but they show an abundant I^+ peak at m/e 127, and may show an $M - H_2I$ peak. Iodine is mono-isotopic, therefore the characteristic doublets shown by Cl and Br ions are not observed in analogous iodine ions.

Aryl halides show abundant molecular ion peaks, but most of the other abundant ions are associated with fragmentation of the aryl cation.

Acyl halides of the aliphatic series may show fragmentations characteristic of alkyl halides and of carbonyl compounds; thus in a variety of cases peaks are found corresponding to HCl^+, $M - 36/38$, $COCl^+$ (m/e 63 and 65), RCO^+, etc. The remaining fragmentations are associated with rupture of the alkyl chain.

Aryl acid chlorides fragment almost entirely by loss of $Cl^·$ from the molecular ion; thereafter the well established decay of $ArCO^+$ takes place as discussed in 7.6.

7.21   Mercaptans, Sulphides, and Thioacids

These fragment similarly to their oxygen analogues (alcohols or phenols, ethers, carboxylic acids), but one notable difference is due to the presence of 4 per cent of ^{34}S: abundant peaks containing sulphur will show the associated peak 2 mass units higher, so that an $M + 2$ peak will appear as 4 per cent of M^+.

Aliphatic thiols may show peaks for $M - H_2S$ ($M - 34$), and also for S^+, HS^+, and H_2S^+ (m/e 32, 33, and 34 respectively).

The thiophene ring gives an abundant molecular ion, which then ruptures to eliminate the thioformyl group HCS; two peaks arise from this process, either at m/e 45 (HCS^+) or $M - HCS$, which arises from charge retention by the remaining fragment. Elimination of $CH{\equiv}CH$ also occurs (reminiscent of Ph^+) which in thiophene itself gives a peak at m/e 58 ($M - 26$). Substituents on the thiophene ring will affect the importance of these fragmentations.

8

The Chemistry of the Class Reactions

It would not be appropriate in a laboratory text to discuss all the chemical reactions used, but this chapter highlights a few of the points of chemical interest in the analysis. These points are either mechanistic interpretations of fundamental reaction types, or else mere commentary on the course of the less well known reactions.

Few details are given on the mechanisms of standard reactions (e.g., the addition of bromine to alkenes).

8.1 Lassaigne's Sodium Fusion Test

The actual fusion process is an extremely rough chemical reaction, and detailed knowledge of the processes involved is scanty; as indicated in 1.4, the main products to be detected are the following ions: CN^-, S^{2-}, halide, and PO_4^{3-}.

Cyanide ion is converted to Prussian blue, which is usually represented as $Fe_3[Fe(CN)_6]_4$. In the solid state, Prussian blue is $KFe(III)[Fe(II)(CN)_6]$, but in solution the Fe oxidation states are reversed. (See L. D. Hansen et al., *J. Chem. Ed.*, 1969, **46**, 46.)

If a red colour is produced in this test, it is due to iron (III) thiocyanate; thiocyanate ion (CNS^-) is only produced if nitrogen and sulphur are present in the organic compound, and if an insufficient amount of sodium has been used in the fusion (since CNS^- is decomposed by excess sodium: $CNS^- + 2Na \longrightarrow CN^- + S^{2-} + 2Na^+$). Repeat the fusion with more sodium.

Sulphide ion can be detected in such small quantities by sodium nitro-prusside that it often detects minute quantities of sulphur-containing impurities in the organic compound. Indeed for this purpose it is rather more sensitive than is necessary, and the lead acetate test is recommended.

Halides are conventionally detected by silver nitrate, provided any CN^- and S^{2-} have been removed by acidification and boiling (see 1.4c). The rather elaborate process used to distinguish chloride from bromide from iodide is the only satisfactory method. Simple displacement reactions (using Br_2/water and Cl_2/water) do not give reliable results on the fusion solution. Most of the reactions are familiar inorganic chemistry; the only organic one of interest is the conversion of fluorescein to eosin by bromine:

fluorescein (orange-yellow)   eosin (red)

Phosphate ion is detected conventionally by ammonium molybdate, giving the heteropolyacid salt (ammonium molybdophosphate) whose structure is only partially represented by $(NH_4)_3(PMo_{12}O_{40})$.

8.2 Mobility of Halogen

When halogen-containing organic compounds are treated with silver nitrate, the ease with which they give a silver halide precipitate is related to the ease with which halide ion is produced. Ionic halides (e.g., salts) react immediately; acyl halides are hydrolysed rapidly with water to give halide ion.

Covalent halides undergo slow solvolysis (nucleophilic substitution by solvent), which is best carried out in homogeneous solution; thus *ethanolic* silver nitrate is used, in which the majority are soluble. The reaction may be S_N1 or S_N2, or a mixture of both. The faster S_N1 process is favoured by tertiary, allylic, or benzylic halides, because of the stabilization of the carbonium ion intermediates:

tertiary   allylic   benzylic

Primary halides solvolyse almost entirely by the slower S_N2 process, while secondary halides undergo both reaction types.

Aryl halides do not easily undergo nucleophilic substitution unless the intermediate anion is stabilized by electron-withdrawing substituents (e.g., nitro).

8.3   The Solubility Tests. Acidity and Basicity

We must distinguish in these reactions between two different crude concepts of an 'acidic' substance, and we will make the same distinction for 'basic' substances.

The most clearly acidic substances ionize sufficiently in water to give H^+, which changes the colour of indicator paper. Other compounds are too weakly acidic to do this, and yet are strong enough acids to react with a base (hydroxide ion) and in so doing they may produce a soluble salt. For example:

catechol
(not acid to litmus)   soluble in water

We can use more powerful bases (ethoxide ion, amide ion, t-butoxide ion, etc.) and demonstrate degrees of acidity in many more organic compounds such as alkynes, β-diketones, etc. For the purposes of qualitative analysis, we restrict the range of bases to three: (a) water (b) bicarbonate ion and (c) hydroxide ion; these enable us to distinguish fairly strong organic acids (lower aliphatic and aromatic RCO_2H) which react with (a) and (b), from weaker acids (long chain RCO_2H, simple phenols), which only react minimally with (a) and (b), but react fully with (c).

By a similar argument, organic bases can be approximately graded by the strength of the acid with which they react. Thus fairly strong bases (lower aliphatic RNH_2) react with water (weak acid) giving sufficient hydroxide ion to change the colour of indicator paper:

$$CH_3NH_2 + H_2O \rightleftharpoons CH_3\overset{+}{N}H_3 + OH^-$$

Most other amines react with stronger acid (dilute hydrochloric acid) to give salts which may or may not be water soluble:

$$RNH_2 + H^+ + Cl^- \longrightarrow RNH_3^+ + Cl^-$$

Very weak bases (triaryl amines) only react with very strong acid (concentrated sulphuric acid).

It is interesting to note that the extremely strong acid, perchloric acid, is used when titrating amines.

8.4   Action of 2,4-Dinitrophenylhydrazine

The mechanism for this reaction is shown in chapter 4, 4.2a.

If the test is carried out with insufficient acid present, then an apparent positive reaction will be given by organic amines. The reagent is in solution as its salt (sulphate), which is more soluble in methanol than the free base. But if an amine is added, it will effectively compete for the sulphuric acid, and free 2,4-dinitrophenylhydrazine may crystallize out, giving the appearance of a positive test.

There should be no difficulty in distinguishing 2,4-dinitrophenylhydrazine: it melts at 194° and is soluble in concentrated hydrochloric acid.

Positive 2,4-D.N.P. tests are given by imines (Schiff's bases) since they are hydrolysed in acid media to an aldehyde and an amine. Bisulphite compounds of aldehydes and ketones also react because of hydrolysis.

8.5   Reducing Properties

If an organic compound is easily oxidized, then we must say also that it possesses reducing properties. For convenience, we restrict the range of oxidizing agents so that we can more easily relate positive tests in these reactions to the presence of particular functions in the molecule. Potassium permanganate (1.11), Tollens' reagent (1.12a), and Fehling's solution (1.12a) are the usual oxidizing agents chosen.

Permanganate ion is the most powerful, and will react with (and be decolourized by) alkenes, aldehydes, α-diketones, etc. Alkenes are oxidized to *cis*-glycols or *erythro*-glycols, aldehydes to carboxylic acids; α-diketones undergo cleavage at the —CO—CO— bond, and α-hydroxy ketones are first oxidized to diketones, which then undergo cleavage.

In Tollens' reagent, the oxidizing agent is Ag^+ (as $Ag(NH_3)_2{}^+$), and in Fehling's solution it is Cu^{2+}: these are only capable of reaction with *most* aldehydes (not all), and with certain polyhydric phenols and amines.

Schiff's reagent is a solution of the purple triphenylmethane dyestuff rosaniline, which has been reduced by SO_2. This is a reversible process; it is reversed by the addition of aldehyde, which forms the bisulphite compound with SO_2, thus removing SO_2 from the equilibrium with rosaniline. The purple colour returns.

8.6   Hydroxamic Acids

Carboxylic esters react with hydroxylamine:

Anhydrides and amides react analogously. Ferric ion forms coloured ligand complexes with the hydroxamic acid; in these the electrons are delocalized over the Fe(III) atoms and the organic moiety.

8.7   Molisch's Test

When carbohydrates are treated with acids, they are dehydrated to furan derivatives: pentoses give furan-2-carboxaldehyde (furfural), and hexoses give the hydroxymethyl derivative:

These aldehydes condense with phenols (e.g., 1-naphthol) to give highly coloured materials of uncertain composition.

8.8   Reactions of Alcohols

Lucas' test depends on the ease of nucleophilic displacement of OH by Cl^-; the OH is first protonated, and comes off as water (good leaving group). The arguments are the same as those outlined in 8.2.

Acetyl chloride reacts with tertiary alcohols to give esters, but the acetate group breaks away (good leaving group) from the tertiary carbonium ion (relatively stable, see 8.2) which then condenses with chloride ion:

$$R_3COH \longrightarrow R_3CO\underset{\underset{O}{\|}}{C}CH_3 \longrightarrow {}^-O\underset{\underset{O}{\|}}{C}CH_3 + R_3C^+ \xrightarrow{\ Cl^-\ } R_3CCl$$

8.9   Ferric Chloride Test for Phenols and Enols

The particular colour produced in the reaction between Fe(III) and a phenol (or enol) cannot be used confidently to identify the phenol (or enol). Colours are produced because of complex formation; this is similar to the hydroxamic acid colours with Fe(III) mentioned in 8.6.

8.10   Action of Nitrous Acid on Amines

All primary and secondary amines react with HNO_2 giving initially the N-nitroso amine, but subsequent decomposition differs in the following ways:

Primary amines: the N-nitroso compound (R.NH.NO) isomerizes to R.N=N.OH, which, after protonation, splits out water to leave R.N$\overset{+}{=}$N. This diazonium ion may be stabilized (i.e., if R is an aromatic ring) or it may immediately lose nitrogen (if R is alkyl or $ArCH_2$, etc.). o-Diamines undergo intramolecular coupling:

Secondary amines: the N-nitroso compound usually isomerizes in acid to the *p*-nitroso compound.

Tertiary amines have already been discussed in 4.3q.

8.11   Amino Acids

The amino group attached to an aromatic ring is only weakly basic because of lone pair donation to the ring; such amino groups are not basic enough to abstract a proton from carboxyl, and therefore amino acids like anthranilic acid (I) exist with free NH_2 and free CO_2H.

Alkyl amino groups are more basic, and amino acids with such groups exist as zwitterions (II).

Separation of Mixtures

Before attempting to separate the constituents of a mixture of organic compounds, it is worthwhile carrying out the tests given in chapter 1 on the mixture. An infrared spectrum should also be recorded. These will allow you to deduce the presence of functional groups, and this information will help in deciding the best method of separating the components.

No general method is available, but most mixtures can be separated by one of the following outline procedures; where possible make use of chemical means.

Chemical Means

In all cases, partition may be improved by the use of a solvent such as ether or chloroform.

(a) **Basicity.** Most amines can be extracted with much dilute hydrochloric acid.

(b) **Acidity.** Acids and phenols can be extracted with dilute sodium hydroxide solution. Acids alone can be extracted with bicarbonate, enabling acids and phenols to be separated. (Phenols can be precipitated from sodium hydroxide solution by passing in CO_2.)

(c) **Carbonyl Compounds** form bisulphite compounds with *saturated* sodium bisulphite solution (the bisulphite compound may crystallize out or stay in the bisulphite solution). The aldehyde or ketone is regenerated on treatment with strong sodium carbonate solution.

(d) **Primary, Secondary, and Tertiary Amines.** (The Hinsberg separation.) Tosyl derivatives (see 4.3a) of primary amines are alkali soluble; those of secondary amines are not. Tertiary amines do not react with toluene-*p*-sulphonyl chloride, and can be steam distilled from primary or secondary amine tosyl derivatives.

Physical Means

The following methods are arranged in an approximate order of increasing efficiency or utility.

(a) **Distillation** of liquid mixtures requires very good fractionation to be useful; use distillation as a last resort unless the b.p. of the components are grossly different, e.g., 50° apart.

(b) **Steam Distillation** is of limited value unless the efficiency can be improved if one of the components forms a salt (i.e., is an acid or a base, in which case steam distil from alkali or acid solution respectively).

(c) **Solvent Separations.** In general, the more highly polar a compound, the less soluble it will be in non-polar solvents (ether, benzene, etc.). For separation to be effective by differential solubility, very big differences in polarity must exist between the components. For example, carbohydrates and hydrocarbons are easily separated by extraction with ether. Use only very polar solvents (water) or 'completely' non-polar solvents (ether, benzene, light petroleum).

(d) **Chromatography** is very widely applicable. Carry out trial separations on T.L.C. plates (e.g., microscope slides dipped in a slurry of silica gel in chloroform). The bulk of the separation should be carried out on columns of (e.g.) silica gel or alumina. As a rough guide, each gramme of mixture needs 15–30 g of adsorbent. The following solvents should be tried in succession: hexane or light petroleum (b.p. 40–60°), benzene, ether, chloroform, ethanol, or methanol.

Name Tabs

These may be cut out, folded, and glued to the top of the appropriate pages for easy access.

glue   fold   glue

Analysis Summary		
Chemical Examination		
Classification		
Separation of Mixtures		
Derivative Preparation		
Derivative Tables		
Infrared Spectrum		
N.M.R. Spectrum		
Electronic Spectrum		
Mass Spectrum		

Index

PRINTED AND BOUND AT WILLIAM CLOWES & SONS LIMITED, BECCLES AND LONDON